# Sustainable Aviation

The Sustainable Aviation book series focuses on sustainability in aviation, considering all aspects of the field. The books are developed in partnership with the International Sustainable Aviation Research Society (SARES). They include contributed volumes comprising select contributions to international symposiums and conferences, monographs, and professional books focused on all aspects of sustainable aviation. The series aims at publishing state-of-the-art research and development in areas including, but not limited to:

- Green and renewable energy resources and aviation technologies
- Aircraft engine, control systems, production, storage, efficiency, and planning
- Exploring the potential of integrating renewables within airports
- Sustainable infrastructure development under a changing climate
- Training and awareness facilities with aviation sector and social levels
- Teaching and professional development in renewable energy technologies and sustainability

* * *

T. Hikmet Karakoc • Oleksandr Zaporozhets
Alper Dalkiran • Ali Haydar Ercan
Editors

# The Future of Electric Aviation and Artificial Intelligence

## Proceedings of the 2023 International Symposium on Electric Aviation & Autonomous Systems and the International Symposium on Unmanned Systems, AI, Design & Efficiency

Springer

*Editors*
T. Hikmet Karakoc
Faculty of Aeronautics and Astronautics
Eskisehir Technical University
Eskisehir, Eskisehir, Türkiye

Information Technology Research
and Application Center Istanbul
Ticaret University
Istanbul, Türkiye

Oleksandr Zaporozhets
Institute of Aviation
Łukasiewicz Research Network
Warsaw, Poland

Alper Dalkiran
School of Aviation
Süleyman Demirel University
Keciborlu, Isparta, Türkiye

Ali Haydar Ercan
UAV Technology Operator Program
Eskisehir Technical University
Eskisehir, Türkiye

International Symposium on Electric Aircraft and Autonomous Systems, iseas, iseas2023, ISEAS, Warsaw, Poland, 2023-7-5–2023-7-7, https://2023.iseasci.org/

International Symposium on Unmanned Systems and The Defense Industry, isudef, isudef2023, ISUDEF, 2023-6-7–2023-6-9, https://2023.isudef.org/

ISSN 2730-7778 ISSN 2730-7786 (electronic)
Sustainable Aviation
ISBN 978-3-031-62093-5 ISBN 978-3-031-62094-2 (eBook)
https://doi.org/10.1007/978-3-031-62094-2



This Springer imprint is published by the registered company Springer Nature Switzerland AG
The registered company address is: Gewerbestrasse 11, 6330 Cham, Switzerland

# Preface

Conversationally referred to as drones, Urban Air Vehicles (UAVs) have undergone significant advancement, manifesting a paradigm shift across diverse industries. Their newborn utility spans agriculture, infrastructure management, cargo logistics, security surveillance, and entertainment. Prospective trajectories in UAV development denote a trajectory marked by augmented autonomy, AI integration, swarm capabilities, urban mobility solutions, refined delivery systems, and advanced environmental monitoring modalities.

The evolution of UAVs is characterized by a regulated endeavor to augment autonomy, enabling drones to undertake complicated tasks with minimal human intervention. The ascendance of AI-infused UAVs heralds an era where these aerial platforms assert dominance by harnessing machine learning algorithms to perceive and dynamically adapt to environmental stimuli. Concurrently, the conceptualization of drone swarms engenders multifaceted prospects across domains ranging from environmental surveillance and military endeavors to disaster response, leveraging synchronized fleets for nuanced problem-solving in complex scenarios.

Urban air mobility emerges as a transformative vision underscored by UAVs, offering a salient solution to burgeoning urban mobility challenges. The envisaged role of drones as efficient aerial conveyances holds the potential to revolutionize urban transportation paradigms, mitigating traffic congestion and fostering heightened productivity. Moreover, UAVs are poised to play a pivotal role as "Monitoring Agents of Environmental Abuses," leveraging their aerial capabilities to gather crucial data on climate change, air quality, and natural disasters, contributing to environmental conservation and disaster management efforts.

As the UAV landscape matures, scholarly discourse is increasingly directed toward fostering sustainable energy solutions to propel these aerial platforms. Novel fuel and propulsion technologies are under scholarly scrutiny, poised to supplant conventional fossil fuel dependencies with ecologically sound alternatives, thereby engendering environmentally conscientious operational frameworks. Concurrently, additive manufacturing techniques and topological optimization methodologies are enlisted to refine aircraft design, curtail weight, and bolster energy efficiency. Despite prevailing challenges, such as the constrained energy storage capacities inherent to electrical aircraft, ongoing academic inquiry and technological innovation collectively proffer auspicious pathways toward materializing thriving regional aircraft fleets and attaining the zenith of zero-emission aviation.

Given the many potential applications and ongoing technological advancements described above, the prospects for autonomous UAVs appear more promising to explore and develop. As technological capabilities evolve, we envisage the emergence of UAVs characterized by enhanced efficacy, reliability, and adaptability. In this manner, International Symposium on Unmanned Systems: AI, Design & Effeciency (ISUDEF) '23, an international and multi-disciplinary symposium, was held online between 7 and 9 June 2023 to address current issues in Research Developments in UAVs in Stockholm, Sweden.

In addition to this very proceedings book, we must say something specific to the "electric aircraft" phenomenon, although those are very close subjects. Electric aircraft's limited energy storage capacity impedes the widespread adoption of regional aircraft fleets and the realization of zero-emission flight. Overcoming this challenge has become a focal point of recent academic research and technological development in the aviation industry. Addressing the energy storage constraints of electric aircraft involves multifaceted approaches, ranging from advancements in battery technology to novel hybrid propulsion systems. Thus, International Symposium on Electric Aviation and Autonomous Systems (ISEAS) '23, an international and multi-disciplinary symposium on electric aviation and autonomous solutions, was held online between 5 and 7 July 2023 in Warsaw, Poland, to help solve the above problems and address electric aircraft systems and safe, reliable electric power in aviation.

We have kindly invited academics, scientists, engineers, practitioners, policymakers, and students to attend the two International Sustainable Aviation and Energy Research Society (SARES) Symposiums of ISUDEF and ISEAS symposiums to share knowledge, explain new technologies and discoveries, and consider the future direction, strategies, and goals in the autonomy, AI, and swarm power of distributed actions. This conference featured keynote presentations by invited speakers and general papers in oral and poster sessions.

There are six main categories found in presented papers at the ISUDEF '23 and ISEAS '23 Symposiums: "UAV Technologies and Operations," "Alternative Energy Sources and Sustainability," "Air Traffic Management and Capacity," "Technological Advancements in Aviation," "Hydrogen-Powered Aircraft and Infrastructure," and "Hybrid Propulsion and Noise Reduction." Those groups have the following key topics:

- UAV Technologies and Operations:
  - UAV Design
  - AI Drone Operation
  - UAV Energy Consumption

  - UAV Sensor Faults
  - Aviation Operational sustainability
- Alternative Energy Sources and Sustainability:
  - Fuel Cell Exergy
  - UAV Sustainability Monitoring
  - Sustainable Aviation Fuels
  - Electric Aircraft Battery
  - Fuel-Cell Aircrafts
  - Hybrid Powered Aircrafts
  - Supersonic Aircraft Emissions
- Air Traffic Management and Capacity:
  - Air Traffic Capacity Assessment Simulation
- Technological Advancements in Aviation:
  - Thermal Turbulent Simulation
  - Electronic Equipment Electromagnetic Compatibility
  - Smart Energy for Airports
  - Fuel burn emissions
- Hydrogen-Powered Aircraft and Infrastructure:
  - Commuter Aircraft Hydrogen Storage
  - Liquid Hydrogen Tank Design
- Hybrid Propulsion and Noise Reduction:
  - Hybrid Propulsion of Aircraft
  - Noise Characteristics of Turboprop Engines
  - Noise Filtering
  - Accurate Noise Assessment

Integrating cutting-edge technologies and sustainable practices has become paramount in today's rapidly evolving aviation landscape. UAVs have emerged as versatile tools with applications ranging from aerial surveillance to package delivery, driving innovations in UAV design and operation. Advanced AI algorithms now enable autonomous drone operations, optimizing flight paths and tasks while minimizing human intervention. However, ensuring the reliability and efficiency of UAVs requires addressing challenges such as energy consumption and sensor faults and driving research into sustainable energy sources like fuel cells and electric batteries. Moreover, noise filtering techniques are essential to mitigate the environmental impact of UAV operations, contributing to the overall sustainability of aviation activities.

As the aviation industry seeks to reduce its environmental footprint and enhance operational efficiency, alternative energy sources have garnered significant attention. Fuel cell technology holds promise for powering aircraft, offering high energy density and zero-emission operation. Efforts to monitor and improve the sustainability of UAV operations encompass developing innovative monitoring systems and utilizing sustainable aviation fuels. Moreover, the emergence of electric aircraft batteries and hybrid propulsion systems further accelerates the transition toward greener aviation. However, successfully integrating these technologies relies on comprehensive assessments of air traffic capacity and noise characteristics, ensuring compatibility with existing infrastructure and regulatory frameworks. The aviation industry can achieve sustainable growth while minimizing environmental impact by advancing research.

We want to thank Springer's editorial team for their support toward the preparation of this book and the chapter authors and reviewers for their outstanding efforts.

On the other hand, we would like to give special thanks to the SARES Editorial office members, who are the heroes behind the veil of the stage, for gathering these chapters. Elif Karakılıç has played a significant role in sharing the load and managing the chapters with Onur Cömert. We thank them for their great effort in the long run for symposium author communications.

Eskisehir, Türkiye — T. Hikmet Karakoc
Warsaw, Poland — Oleksandr Zaporozhets
Keciborlu, Isparta, Türkiye — Alper Dalkiran
Eskisehir, Türkiye — Ali Haydar Ercan

# Contents

# Chapter 1
# Parametric Approach to Initial Weight Determination in the Preliminary Design of a Quadrotor Cargo UAV

**Fikret Kamil Corbaci and Yunus Emre Dogan**

## Nomenclature

| | |
|---|---|
| ESC | Electronic speed controller |
| GTOW | Gross take of weight |
| UAV | Unmanned aerial vehicle |

## 1.1 Introduction

Drones could save a lot of manpower and shift unnecessary road traffic to the sky (Elouarouar and Medromi 2022). In the military field, relatively passive missions such as reconnaissance surveillance, in which large, manned aircraft were used in the past, have started to be performed mostly with unmanned aerial vehicles (UAVs), thanks to their ability to perform challenging and hazardous tasks with high mobility, safety, and low cost (Ahirwar et al. 2019). It is possible to find a wide variety of applications, such as meteorological applications, photography, air transport, air cargo transport, forest fire detection and response, agriculture, and search rescue (Jeongeun et al. 2019). The transportation industry is the leader of the pack when it comes to drone use (Elouarouar and Medromi 2022). It refers to the delivery of parcels to the customers and is the most expensive process in distribution logistics (Pugliese et al. 2020). UAV express is a service that allows a light load to be transported in a short period of time in a local area. As the demand for UAVs increased, it became necessary to reduce design times in order to reach different

F. K. Corbaci (✉)
STM A.S., Production Department, Ankara, Turkiye

Y. E. Dogan
Ankara Yildirim Beyazit University, Mechanical Engineering Department, Ankara, Turkiye

 
T. H. Karakoc et al. (eds.), *The Future of Electric Aviation and Artificial Intelligence*, Sustainable Aviation, https://doi.org/10.1007/978-3-031-62094-2_1

types in a short time. In addition to the work done to shorten the design time to get a new quadrotor, this study focused on determining the initial gross takeoff weight.

## 1.2 Design Methodology

With the aim of having an optimum cargo UAV, preliminary design requirements based on mission profile should be driven during the design process. In order to get optimized design solutions, flight time inefficiency, losses in payload capacity, time and cost losses, etc., are also taken into account and the design process should be carried out step by step.

Because it was preferred as the main design optimization approach, the parametric design approach stated by Ullman (2010) was applied during the preliminary phase. At the beginning of the study, the step of determining the necessary parameters was carried out and thrust and maximum continuous current, battery capacity, propeller diameter, and frame size were selected as the basic parameters on which the system features of the cargo UAV depend.

## 1.3 Parametric Approach to Initial Weight Determination of UAV

Techniques and tools for the preliminary design of a quadrotor will be examined within the product design stages of a unique rotary wing unmanned aerial vehicle (quadrotor) that meets the specific requirements. In the preliminary design phase, after the parameters of the final product were determined, the related systems, its subsystems, and components were clearly defined as shown in Fig. 1.1 (Winslow et al. 2018).

Related component requirements were obtained from mission requirements, and design features that meet these requirements were created compatibly.

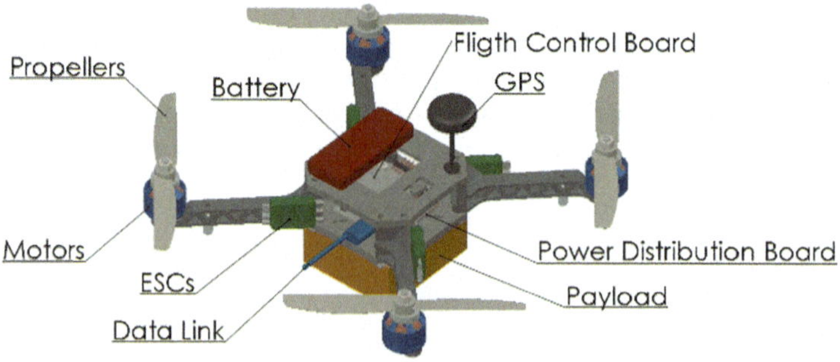


**Fig. 1.1** Fundamental subsystems of a quadrotor UAV

### 1.3.1 Mission Profile and Mission Requirements

The main objective of UAV is to be used to carry lightweight urgent documents among close locations. The load to be carried by the vehicle consists of authentic documents that have a maximum of 200 g weight. It is assumed that the drone takes off, climbs to the lowest altitude of 40 m, and moves at cruising speed toward the delivery point. With the baseline of these mission profiles, the minimum performance requirements were determined as shown in Table 1.1.

It is assumed that the quadrotor UAV to be discussed in the study will be made of ABS material, consist of two plates, upper and lower, and will have a true-X geometry.

### 1.3.2 Regression Analysis of GTOW for Parametric Design

In the preliminary design of the cargo UAV, the masses of the components should be calculated primarily in order to see that it can perform the expected performance in the targeted mission specifications. One of the most important design inputs to determine the masses of the components at this stage is the cargo UAV gross takeoff weight (GTOW). It is assumed that GTOW consists of the masses of the engine ($m_{\mathrm{m}}$), ESC ($m_{\mathrm{E}}$), battery ($m_{\mathrm{B}}$), propeller ($m_{\mathrm{p}}$), frame ($m_{\mathrm{f}}$), avionics ($m_{\mathrm{A}}$), and payload ($m_{\mathrm{PL}}$), which are the main components in the UAV.

$$\mathrm{GTOW} = \left(m_{\mathrm{m}} + m_{\mathrm{E}} + m_{\mathrm{B}} + m_{\mathrm{p}} + m_{\mathrm{f}} + m_{\mathrm{A}} + m_{\mathrm{PL}}\right) \cdot g \tag{1.1}$$

In the preliminary design, it is necessary to determine the initial GTOW that will be created by the preliminary estimate of the mass to be revealed for each of the components. All relevant component masses were separately determined with respect to predetermined parameters for each component by using regression analyses. While doing this, the tables prepared for the product alternatives that are sold commercially for each component and containing catalog data about both mass and the relevant parameter were used.

**Table 1.1** Minimum performance requirements

| No | Requirement | Value |
|---|---|---|
| 1 | Average cruise velocity | 5 m/s |
| 2 | GTOW | 12.5 N |
| 3 | Minimum flight time | 10 min |
| 4 | Payload capacity | 0.200 kg |
| 5 | Communication range | 1500 m |
| 6 | Mass of avionics | 0.200 kg |

The equations that emerged after the regression analyses were formed as follows:

- Mass of motor ($m_m$) with respect to the maximum thrust of motor ($T$):

$$m_m = 1E - 07(T)^3 - 0.0003(T)^2 + 0.2783(T) - 56.133 \tag{1.2}$$

- Mass of ESC ($m_E$) with respect to the maximum continuous current of ESC ($I_{max}$) (Vu et al. 2019):

$$m_E = 0.9546(I_{max}) - 2.592 \tag{1.3}$$

- Mass of the battery ($m_B$) with respect to the capacity of the battery ($C$) and number of cells (Vu et al. 2019):

$$\text{For } n = 3 \quad m_B = 0.0625(C) + 35.526 \tag{1.4}$$

$$\text{For } n = 4 \quad m_B = 0.761(C) + 69.522 \tag{1.5}$$

$$\text{For } n = 6 \quad m_B = 0.1169(C) + 132.0 \tag{1.6}$$

- Mass of the propeller ($m_p$) with respect to the diameter of the propeller ($d_p$) (Vu et al. 2019):

$$m_p = 0.1367(d_p)^2 - 9.317(d_p) + 0.881 \tag{1.7}$$

- Mass of the frame ($m_f$) with respect to the diagonal size of the frame ($l$) and thickness ($t$):

$$\text{For } t = 3 \text{ mm}, \quad m_f = 3E - 06(l)^3 - 0.003(l)^2 + 1.3666(l) - 106.53 \tag{1.8}$$

$$\text{For } t = 4 \text{ mm}, \quad m_f = 4E - 06(l)^3 - 0.004(l)^2 + 1.8221(l) - 142.04 \tag{1.9}$$

Thus, initial GTOW was determined as a function of all selected parameters:

$$\text{GTOW} = f(T, I_{max}, C, n, \ d_p, l, t) \tag{1.10}$$

This initial GTOW found will be used in the preliminary design of all components, and the necessary selections will be easily completed.

## 1.4 Conclusion

In this study, the equations were created using the regression analysis between the parameters selected for each component and the component masses. With the approach introduced:

- The initial estimation time of GTOW in the preliminary design phase was shortened.
- In addition, the number of iterations performed to achieve GTOW results with acceptable accuracy throughout the same process has been reduced and the process speed has been increased.

We aimed to carry out CFD analyses and examine the performance of the cargo UAV, whose design was completed as a result of continuing the detailed design with the methodology set forth in this study. With the structural optimization to be carried out during the detail design phase, it is targeted to reduce weight and, accordingly, reduce the product cost. Also, the weight values determined according to the approach in this study will be compared with the final values that will emerge after the design of the cargo UAV is completed and the effectiveness of the approach will be evaluated.

## References

Elouarouar, S. and Medromi, H. (2022). Multi-Rotors Unmanned Aerial Vehicles Power Supply and Energy Management. E3S Web of Conferences, 336, p. 00068. https://doi.org/10.1051/e3sconf/202233600068

Ahirwar, S., Swarnkar, R., Bhukya, S. and Namwade, G. (2019). Application of drone in agriculture. International Journal of Current Microbiology and Applied Sciences, 8(01), pp. 2500–2505. https://doi.org/10.20546/ijcmas.2019.801.264

Jeongeun, K., Seungwon, K., Chanyoung, J. and Hyoungil, S. (2019). Unmanned Aerial Vehicles in Agriculture: A Review of Perspective of Platform, Control, and Applications. IEEE Transactions and Journals, Volume 4, 2016, pp. 1–17. https://doi.org/10.1109/ACCESS.2019.2932119

Pugliese, L.D.P., Guerriero, F. and Macrina, G. (2020). Using drones for parcels delivery process. Procedia Manufacturing, 42, pp. 488–497. https://doi.org/10.1016/j.promfg.2020.02.043

Ullman D., (2010). The Mechanical Design Process, Fourth Edition, The McGraw-Hill Companies, Inc., p. 36.

Winslow J, Hrishikeshavan V, and Chopra I (2018) Design methodology for small-scale unmanned quadrotors. Journal of Aircraft, 55(3):1062–1070. https://doi.org/10.2514/1.C034483

Vu NA, Dang DK, Dinh TL (2019) Electric propulsion system sizing methodology for an agriculture multicopter. Aerospace Science and Technology, 90:314–326. https://doi.org/10.1016/j.ast.2019.04.044

# Chapter 2
# Comparative Analysis of AI-Supported Drone Operations: Evaluating Critical Aspects Among Pilot Groups

**Omar Alharasees and Utku Kale**

## Nomenclature

| | |
|---|---|
| AHP | Analytical hierarchal process |
| AI | Artificial intelligence |
| OODA | Observe–orient–decide–act |
| UAS | Unmanned aircraft system |

## 2.1 Introduction

The integration of artificial intelligence (AI) into drone operations dates back to the Cold War, initially providing limited autonomous capabilities and following pre-programmed flight paths (Wu et al. 2022). Over time advancements in AI technology have significantly increased drones' autonomy (Li et al. 2022), enabling real-time decision-making and enhancing efficiency, versatility, and precision in various domains such as surveillance, delivery, and search and rescue missions. The observe–orient–decide–act (OODA) loop, a decision-making framework developed by John Boyd, has been gradually adopted in drone operations, offering a valuable tool for navigating complex and dynamic situations (Johnson 2022). Its integration has evolved progressively, starting from its conceptual inception and gaining widespread acceptance. The OODA loop framework is integral to drones' ability to perceive, analyze, and adapt to real-time environmental changes (Osinga 2007). Integrating the OODA loop enhances drone pilots' decision-making capacities,

O. Alharasees (✉) · U. Kale
Department of Aeronautics, and Naval Architecture, Faculty of Transportation Engineering and Vehicle Engineering, Budapest University of Technology and Economics, Budapest, Hungary
e-mail: oalharasees@edu.bme.hu; kale.utku@kjk.bme.hu



T. H. Karakoc et al. (eds.), *The Future of Electric Aviation and Artificial Intelligence*, Sustainable Aviation, https://doi.org/10.1007/978-3-031-62094-2_2

expediting their responses (Zhou et al. 2020). As AI and drone operations advance, the OODA loop becomes increasingly essential for seamless drone operations in complex contexts.

The future application of the OODA loop in drones is set to expand, leveraging AI advancements. This expansion will enhance the efficiency and decision-making abilities of drone pilots. The structured nature of the OODA loop will continue to be a cornerstone in the evolution of AI-driven drone operations, enabling real-time adaptation to changing circumstances. AI is a key component in advancing drone operations, emphasizing cognitive capabilities, autonomy, and safety. Future advancements in AI-driven drones will offer refined situational awareness, decision-making prowess, and seamless communication (Al-Turjman and Zahmatkesh 2020). As drones are increasingly deployed across industries, the role of AI in ensuring secure and efficient operations becomes paramount.

AHP stands as a comprehensive multi-criteria decision-making (MCDM) technique that provides a well-structured framework for evaluating intricate decision problems (Saaty 2008; Saaty and Vargas 2012). AHP has been effectively employed in numerous studies concerning the evaluation of drone operations. Its applications span a wide spectrum, encompassing assessments of drone training systems (Ting et al. 2018), drone route optimization (Li et al. 2017), live video streaming implementations (Khanmohammadi et al. 2021), as well as evaluations of human factor involvement (Kale et al. 2021; Alharasees et al. 2022) and drone-assisted disaster response initiatives with a focal point on safety and security (Makalesi et al. 2020). In the context of AI-driven drone operations, AHP emerges as a proficient tool for evaluating the performance of AI integration (Sridevi et al. 2022). Through a meticulous weighing of critical facets like perception, decision-making, and adaptation, AHP enables the derivation of a comprehensive assessment of the system at hand (Ramirez-Atencia et al. 2020).

Employing the analytic hierarchy process (AHP) for AI-driven drone system evaluation provides a comprehensive and unbiased methodology. It ensures a systematic approach for identifying areas of enhancement in AI integration, thus guiding technological advancements. The outcomes of this evaluation significantly influence decision-making and refine AI technologies in the drone domain.

The assimilation of AHP in assessing AI-infused drone systems offers a notable contribution. It establishes a comprehensive evaluative approach, ensuring continual progress and refinement in AI-driven drone operations. By integrating AHP seamlessly, a methodical framework emerges for assessing pivotal aspects, informing technological advancements and enhancing AI integration performance within drone operations.

## 2.2 Method

The integration of AI into drone and unmanned aircraft system (UAS) operations signifies a significant leap forward in aviation. As the complexity and quantity of drones and UASs escalate, the demand for advanced AI-supported systems becomes increasingly evident. These systems are tailored to optimize decision-making, enhance operational efficiency, and augment situational awareness, ultimately heightening the safety and efficacy of unmanned operations.

This study adopts a methodology that combines the classical OODA loop with the AHP method to evaluate AI-drone systems. The OODA loop, as depicted in Fig. 2.1, serves as a decision-making framework that enables timely comprehension and response to situations. By incorporating the OODA loop within the AHP framework, this study endeavors to establish a hierarchical structure encompassing key components of AI-drone operations—perception, decision-making, and adaptation—weighted according to their significance, as illustrated in Fig. 2.2.

The resulting hierarchy engenders a comprehensive evaluation model for AI-drone operations. Employing the AHP methodology enables a comprehensive assessment of the present state of AI-drone operations, identification of areas necessitating improvement, and guidance for the development of novel technologies. The integration of AHP within the OODA loop framework provides crucial insights into the incorporation of AI in drone operations and facilitates the creation of a hierarchical AI-drone OODA model. AI-driven drones rely on the OODA loop (observe, orient, decide, act) for their functionality. It encompasses essential elements of data collection, assessment, decision-making, and action execution. The loop can be further divided into subcomponents specific to AI-supported drones. Observation involves collecting sensor data (e.g., cameras, lidar) and perception to understand the environment and maintain situational awareness. Orientation assesses data relevance, contextualizes information based on mission and

**Fig. 2.1** OODA loop

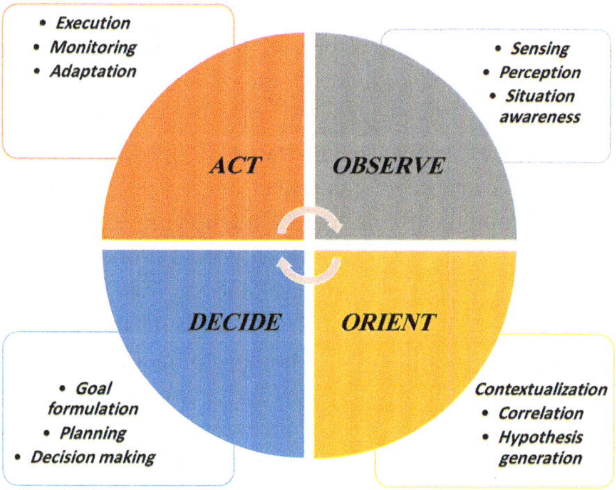

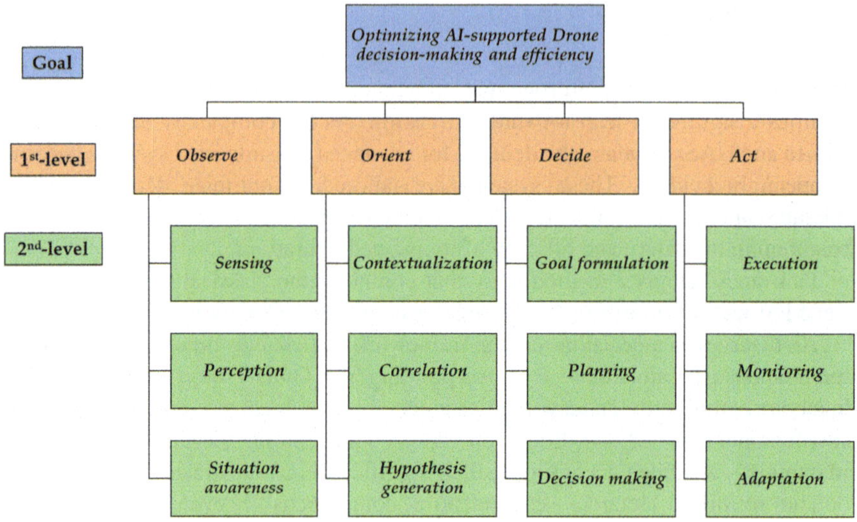


**Fig. 2.2** The hierarchal model

**Table 2.1** Saaty scale (Saaty 1977)

| Mathematical representation | 1 | 2, 4, 6, 8 | 3 | 5 | 7 | 9 |
|---|---|---|---|---|---|---|
| Meaning | Both are equally important | Values between moderate and strong importance | One is moderately more important than the other | One is strongly more important than the other | One is very strongly more important | One is extremely more important than the other |

capabilities, and generates hypotheses to explain the situation. Decision-making encompasses goal formulation, planning, and selecting the best course of action considering available information and constraints.

Action involves executing the chosen course of action, monitoring performance, and adapting to changes in the environment or mission goals. In evaluating AI-supported drones using the AHP methodology, the OODA loop serves as a criterion. Pairwise comparisons assess the relative importance of its components and subcomponents, aiding in identifying areas for improvement and guiding the development of advanced AI-supported drone systems. AHP utilizes pairwise comparison matrices (PCM) to evaluate the importance of options. Decisions are based on comparisons using the Saaty scale (Table 2.1). Geometric mean calculations determine the influence of model components, aiding prioritization. Achieving a matrix consistency ratio below 0.1 is crucial for reliable evaluations, considering the tendency of matrices to be inconsistent. Consistency ratios are computed for the groupings.

## 2.3 Results and Discussion

This study involved an online survey using AHP to gather insights from two distinct groups of experts (drone pilots) divided based on their experience level. The purpose was to compare the perspectives of experienced drone pilots with more than 3 years of operational experience and less-experienced drone pilots with less than 3 years of operational experience. The survey design followed a two-level hierarchical model based on the OODA loop hierarchy. The survey participants consisted of 20 drone pilots from 12 different countries, as illustrated in Fig. 2.3 (right); however, Fig. 2.3 (left) illustrates the participated drone pilots' work field distribution.

Analysis of survey respondents' viewpoints on the evaluation of the OODA hierarchical approach uncovers contrasting perspectives influenced by differing levels of expertise, professional categorizations, and understanding of the present situation and its evolution. Nevertheless, the application AHP for pairwise comparisons offers a comprehensive outlook on the advancement of AI-supported drone operations within the contemporary framework, surpassing simplistic methodologies. Utilizing the prescribed methodology, initial categorization based on experience classifications was performed, and average responses were calculated using the geometric mean. Each group determined weights and consistency ratios for the first level in the OODA model, which are presented in Tables 2.2 and 2.3.

The analysis of the pairwise comparison matrices (PCMs) reveals that experienced drone pilots assign the highest importance to the "observe" and "act" stages of the OODA loop. They recognize the criticality of accurate data collection and effective execution in their decision-making process.

While with less experienced drone pilots, in contrast, the PCMs indicate that less experienced drone pilots prioritize the "act" stage of the OODA loop the most, followed by "decide." They may rely more on taking immediate action and making decisions based on their limited operational experience.

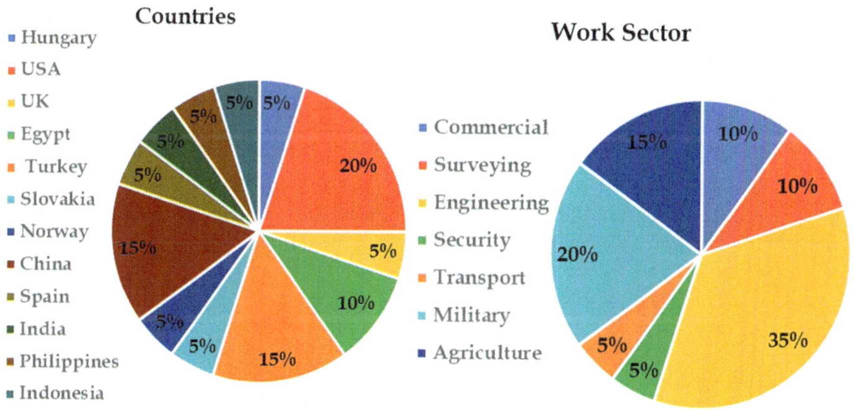


**Fig. 2.3** The participant countries (left) and work sector distribution (right)

**Table 2.2** Experienced drone pilots' group PCM

| Experienced Drone Pilots | | | | | |
|---|---|---|---|---|---|
| OODA MODEL | OBSERVE | ORIENT | DECIDE | ACT | Weights |
| OBSERVE | 1 | 0.62 | 0.91 | 0.62 | 28.22% |
| ORIENT | 1.61 | 1 | 0.59 | 0.34 | 13.11% |
| DECIDE | 1.10 | 1.68 | 1 | 1.79 | 30.62% |
| ACT | 1.60 | 2.93 | 0.56 | 1 | 28.05% |
| CR= 0.086 | Sum= | | | | 100.00% |

**Table 2.3** Less experienced drone pilots' group PCM

| Less Experienced Drone Pilots Group | | | | | |
|---|---|---|---|---|---|
| OODA MODEL | OBSERVE | ORIENT | DECIDE | ACT | Weights |
| OBSERVE | 1 | 0.48 | 0.41 | 0.67 | 19.48% |
| ORIENT | 1.02 | 1 | 0.52 | 0.98 | 18.16% |
| DECIDE | 1.74 | 1.89 | 1 | 1.02 | 30.48% |
| ACT | 1.90 | 2.81 | 0.77 | 1 | 31.88% |
| CR= 0.061 | Sum= | | | | 100.00% |

A comparison between the two groups of drone pilots, as depicted in Fig. 2.4, highlights notable differences in their prioritization within the OODA loop. Experienced drone pilots exhibit a higher emphasis on "observation" and "decision-making" stages, reflecting their advanced skills and comprehensive understanding of the operational environment. On the other hand, less experienced drone pilots place greater importance on the "action" stage, possibly due to their inclination toward immediate response and limited exposure to complex decision-making scenarios. These findings illustrate the impact of experience levels on the perception and prioritization of key OODA loop components among drone pilots.

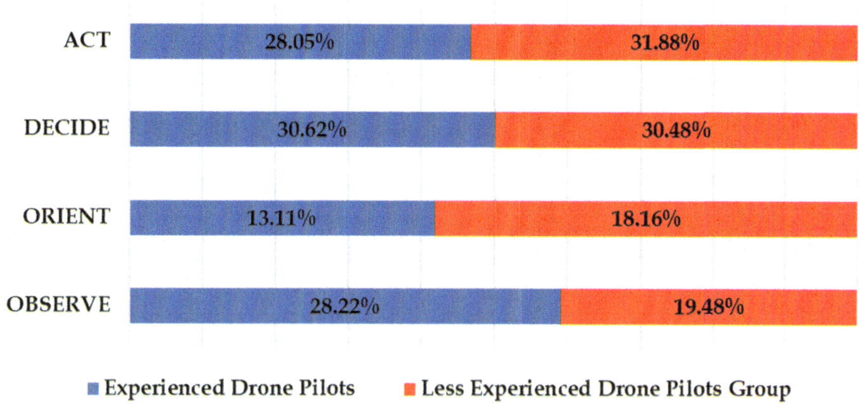


**Fig. 2.4** OODA loop first-level comparison

## 2.4 Conclusion

This study proposes the integration of the observe–orient–decide–act (OODA) loop with the analytic hierarchy process (AHP) to comprehensively evaluate AI-driven drone systems. By considering perception, decision-making, and adaptation, this combined approach facilitates the identification of areas for system improvement. Comparing AHP results across diverse groups of drone pilots emphasizes the operator's role in AI-driven drone operations. The study showcases the value of this integrated approach for evaluating AI-driven drone systems, offering a transparent and systematic means of assessing performance.

The integration of AHP into AI-driven drone system evaluation contributes significantly, providing an objective and comprehensive framework. The evaluation outcomes inform decision-making and guide the development of advanced AI technologies for drones, ensuring continual enhancement of their operations. With ongoing advancements in AI technology, the utilization of the OODA loop in AI-driven drone operations is poised to expand further, augmenting the decision-making capabilities of drone pilots. This combined AHP and OODA loop approach presents a promising solution, offering a structured methodology to evaluate AI-driven drone systems, assess their performance, and pinpoint areas for improvement.

**Acknowledgments** This work was supported by Project No. 2022–2.1.1-NL-2022-00012, which has been implemented with the support provided by the Ministry of Culture and Innovation of Hungary from the National Research, Development and Innovation Fund, financed under the National Laboratories funding scheme.

## References

Al-Turjman, F. and Zahmatkesh, H. (2020) 'A Comprehensive Review on the Use of AI in UAV Communications: Enabling Technologies, Applications, and Challenges', *Unmanned System Technologies*, pp. 1–26. Available at: https://doi.org/10.1007/978-3-030-38712-9_1/COVER.

Alharasees, O., Al-Rabeei, S. and Kale, U. (2022) 'Human Factors Analysis and Classification System – AHP Drone Model Assessment', *Acta Avionica*, 24(2), pp. 41–49. Available at: https://doi.org/10.35116/AA.2022.0030.

Johnson, J. (2022) 'Automating the OODA loop in the age of intelligent machines: reaffirming the role of humans in command-and-control decision-making in the digital age', *Defence Studies* [Preprint]. Available at: https://doi.org/10.1080/14702436.2022.2102486.

Kale, Utku *et al.* (2021) 'Objective Measurement of Human Factors for Supporting the Operator's Load Simulation and Management', in *32nd Congress of the International Council of the Aeronautical Sciences*. Shanghai. Available at: https://www.researchgate.net/publication/359513201_OBJECTIVE_MEASUREMENT_OF_HUMAN_FACTORS_FOR_SUPPORTING_THE_OPERATOR'S_LOAD_SIMULATION_AND_MANAGEMENT (Accessed: 16 April 2022).

Khanmohammadi, E., Barekatain, B. and Quintana, A.A. (2021) 'An enhanced AHP-TOPSIS-based clustering algorithm for high-quality live video streaming in flying ad hoc networks', *Journal of Supercomputing*, 77(9), pp. 10664–10698. Available at: https://doi.org/10.1007/S11227-021-03645-3/TABLES/6.

Li, S., Cummings, M.L. and Welton, B. (2022) 'Assessing the impact of autonomy and overconfidence in UAV first-person view training', *Applied Ergonomics*, 98, p. 103580. Available at: https://doi.org/10.1016/J.APERGO.2021.103580.

Li, X. *et al.* (2017) 'UAV route evaluation algorithm based on CSA-AHP and TOPSIS', *2017 IEEE International Conference on Information and Automation, ICIA 2017*, pp. 915–920. Available at: https://doi.org/10.1109/ICINFA.2017.8079033.

Makalesi, A., Yıldızbaşı, A. and Gür, L. (2020) 'A decision support model for unmanned aerial vehicles assisted disaster response using AHP-TOPSIS method', *Avrupa Bilim ve Teknoloji Dergisi*, (20), pp. 56–66. Available at: https://doi.org/10.31590/EJOSAT.737764.

Osinga, F.P.B. (2007) *Science, strategy and war: The strategic theory of John Boyd*. 1st edn, *Science, Strategy and War: The Strategic Theory of John Boyd*. 1st edn. London: Routledge Taylor & Francis Group. Available at: https://doi.org/10.4324/9780203088869/SCIENCE-STRATEGY-WAR-FRANS-OSINGA.

Ramirez-Atencia, C., Rodriguez-Fernandez, V. and Camacho, D. (2020) 'A revision on multi-criteria decision making methods for multi-UAV mission planning support', *Expert Systems with Applications*, 160, p. 113708. Available at: https://doi.org/10.1016/J.ESWA.2020.113708.

Sridevi, S., Augustine, J. and Ghosh, S.K. (2022) 'An analytic hierarchy process (AHP) based UAV selection mechanism for beyond 5G networks', *2022 IEEE 21st International Symposium on Network Computing and Applications (NCA)*, pp. 27–34. Available at: https://doi.org/10.1109/NCA57778.2022.10013547.

Saaty, T.L. (1977) 'A scaling method for priorities in hierarchical structures', *Journal of Mathematical Psychology*, 15(3). Available at: https://doi.org/10.1016/0022-2496(77)90033-5.

Saaty, T.L. (2008) 'Decision making with the analytic hierarchy process', *International journal of services sciences*, 1(1), pp. 83–98. Available at: https://www.inderscienceonline.com/doi/abs/10.1504/IJSSci.2008.01759 (Accessed: 15 February 2022).

Saaty, T.L. and Vargas, L.G. (2012) 'The possibility of group choice: Pairwise comparisons and merging functions', *Social Choice and Welfare*, 38(3). Available at: https://doi.org/10.1007/s00355-011-0541-6.

Ting, W.T. *et al.* (2018) 'Research on UAV simulation training system based on visual simulation', *Proceedings of 2018 IEEE International Conference on Mechatronics and Automation, ICMA 2018*, pp. 1972–1977. Available at: https://doi.org/10.1109/ICMA.2018.8484640.

Wu, M., Zhu, Q. and Pang, C. (2022) 'Intelligent Warfare : Prospects of Military Development in the Age of AI', *Intelligent Warfare* [Preprint]. Available at: https://doi.org/10.4324/B22974.

Zhou, Y., Rao, B. and Wang, W. (2020) 'UAV swarm intelligence: Recent advances and future trends', *IEEE Access*, 8, pp. 183856–183878. Available at: https://doi.org/10.1109/ACCESS.2020.3028865.

# Chapter 3
# Nontraditional Attitude Filtering with Uncertain Process Noise

**Chingiz Hajiyev and Demet Cilden-Guler**

## 3.1 Introduction

An extended Kalman filter (EKF) can be used for attitude and angular velocity estimation purposes in a spacecraft. One way is to use the nonlinear measurements (magnetometer, sun sensor, etc.) directly in the filter. The nonlinear measurements of reference directions can be used to create EKF for satellite attitude and rate estimates in traditional approaches (Cilden et al. 2017; Hajiyev and Cilden-Guler 2017; Ivanov et al. 2018; Mashtakov et al. 2020; Springmann and Cutler 2014). The nonlinear models of the reference directions serve as the foundation for the measurement models in the filter. Therefore, nonlinear equations are used to link the data and states.

When using a nontraditional approach, the orientation angles are first determined using vector measurements and the corresponding method of single-frame determination of orientations at each recursive step. Then, these attitude angles are utilized directly as measurement inputs for an EKF (Hajiyev et al. 2016; Hajiyev and Bahar 2003; Hajiyev and Cilden-Guler 2017; Mimasu and Van der Ha 2009).

This study shows that the process noise bias and process noise increment-type system changes will change the statistical characteristics of the EKF innovation sequence. The theoretical basics of the Q-adaptive SVD-aided EKF with uncertain process noise mean and covariance are developed and presented. For the purpose of estimating a nanosatellite's attitude, simulations are compared using the adaptive and nonadaptive versions of the nontraditional attitude filter in the presence of process noise bias and process noise increment-type system changes.

C. Hajiyev (✉) · D. Cilden-Guler
Istanbul Technical University, Faculty of Aeronautics and Astronautics, Istanbul, Turkey
e-mail: cingiz@itu.edu.tr; cilden@itu.edu.tr



T. H. Karakoc et al. (eds.), *The Future of Electric Aviation and Artificial Intelligence*, Sustainable Aviation, https://doi.org/10.1007/978-3-031-62094-2_3

The rest of the chapter is structured as follows. In Sect. 3.2, extended Kalman-based estimator aided by SVD is given. In Sect. 3.3, the results and discussion of the simulations are given. Section 3.4 finally summarizes the study.

## 3.2 Satellite Rotational Motion and Attitude Measurement Models

By using the quaternion attitude representation, the satellite's kinematics equation of motion can be expressed as

$$\dot{\boldsymbol{q}}(t) = \frac{1}{2}\Omega(\boldsymbol{\omega}_{\mathrm{BR}}(t))\boldsymbol{q}(t). \tag{3.1}$$

Here, $\boldsymbol{q}$ consists of four attitude parameters in the quaternion, $\boldsymbol{q} = [q_1 \ q_2 \ q_3 \ q_4]^T$. Since the last term is a scalar and the first three terms are vector terms, we may rewrite the quaternion as $\boldsymbol{q} = \begin{bmatrix} \boldsymbol{g}^T & q_4 \end{bmatrix}^T, \boldsymbol{g} = \begin{bmatrix} q_1 & q_2 & q_3 \end{bmatrix}^T$, $\Omega(\boldsymbol{\omega}_{\mathrm{BR}})$ is the skew symmetric matrix whose elements are the components of $\boldsymbol{\omega}_{\mathrm{BR}}$ angular velocity vector of body frame with respect to the reference (orbital) frame.

It is important to specify the body's angular rate vector in relation to the inertial axis frame independently from the angular velocity vector, $\boldsymbol{\omega}_{\mathrm{BI}} = \begin{bmatrix} \omega_x & \omega_y & \omega_z \end{bmatrix}^T$. $\boldsymbol{\omega}_{\mathrm{BI}}$ and $\boldsymbol{\omega}_{\mathrm{BR}}$ can be related via

$$\boldsymbol{\omega}_{\mathrm{BR}} = \boldsymbol{\omega}_{\mathrm{BI}} - A\begin{bmatrix} 0 & -\omega_{\mathrm{o}} & 0 \end{bmatrix}^T. \tag{3.2}$$

Here, $\omega_{\mathrm{o}}$ indicates the satellite's angular orbital velocity, $\mu$ is the gravitational constant, and $r$ is the distance between the satellite's and Earth's centers of mass. In (3.3), $A$ is the attitude matrix and the quaternions are connected to it by

$$A = \left(q_4^2 - |\boldsymbol{g}|^2\right)I_{3\times 3} + 2\boldsymbol{g}\boldsymbol{g}^T - 2q_4[\boldsymbol{g}\times]. \tag{3.3}$$

$I_{3\times 3}$ is the identity matrix with the dimension of $3 \times 3$ and $[\boldsymbol{g}\times]$ is the skew-symmetric matrix whose elements are the components of $\boldsymbol{g}$ vector.

Based on Euler's equations, it is possible to deduce the satellite's dynamic equations

$$J\frac{\boldsymbol{\omega}_{\mathrm{BI}}}{dt} = \boldsymbol{N}_{\mathrm{d}} - \boldsymbol{\omega}_{\mathrm{BI}} \times (J\boldsymbol{\omega}_{\mathrm{BI}}), \tag{3.4}$$

where $J$ is the principal moments of inertia matrix as $J = \mathrm{diag}\,(J_x, J_y, J_z)$ and $\boldsymbol{N}_{\mathrm{d}}$ is the vector of disturbance torque affecting the nanosatellite.

In this study, we outline our technique for a nanosatellite equipped with magnetometers and a Sun sensor, two different types of attitude sensors. We require the unit vectors generated in the reference orbit frame that correspond to the unit vectors measured by the sensors in the spacecraft body frame in order to determine the attitude. It is assumed that the magnetometers and Sun sensors are calibrated against any bias and/or misalignment.

## 3.3 Integration of SVD and EKF for Attitude Estimation

The nontraditional attitude estimation procedure is proposed, which consists of two stages: singular value decomposition (SVD) and EKF. Due to its greater robustness, the SVD approach is chosen as the single-frame method in this instance. The SVD-based attitude determination algorithm is given in Markley and Mortari (2000).

Since discrete-time nonlinear equations are used to construct the EKF, the following equations represent the process model:

$$\mathbf{x}(k+1) = \mathbf{f}(\mathbf{x}(k), k) + \mathbf{w}(k), \tag{3.5}$$

$$\mathbf{y}(k) = \mathbf{H}\mathbf{x}(k) + \mathbf{v}(k). \tag{3.6}$$

Here, $\mathbf{x}(k)$ is the state vector and $\mathbf{y}(k)$ is the measurement vector. Moreover, $\mathbf{w}(k)$ and $\mathbf{v}(k)$ are the process and measurement error noises, which, according to the assumption, are processes with Gaussian white noise with zero mean and a covariance of $\mathbf{Q}(k)$ and, $\mathbf{R}(k)$ respectively, and $\mathbf{H}$ is the measurement matrix of system.

The extended Kalman filter for attitude and rate estimation is given in Appendix A.

## 3.4 Influence of Process Noise Changes on the Innovation

The process noise bias and process noise increment-type system changes will cause a change in the statistical characteristics of the innovation sequence of EKF. In this section, influence of these types of changes to the innovation of EKF is investigated.

### *3.4.1 Influence of Process Noise Increment on the Innovation*

Let the measurements are processed by the EKF (A.1)–(A.7) and a process noise increment occurs at the iterations $k \geq \tau$. Process noise increment can be simulated by multiplying the process noise vector with the diagonal matrix $\mathbf{F}(k)$, where the diagonal elements meet the following condition: $\sigma_{ii}(k) \geq 1$, $\left(i = \overline{1, s}\right)$ for $\forall k \geq \tau$ ($s$ is the dimension of process noise vector). As is clear, for the process noise change in the process channel $j$ the appropriate diagonal element of $\mathbf{F}(k)$ will be larger than

1, that is, $\sigma_{jj}(k) > 1$ for $\forall k \geq \tau$ and the rest of the process channels become 1. Consequently, the diagonal elements of $\mathbf{F}(k)$ can be presented in the following form:

$$\sigma_{ii} = \begin{cases} 1 : \text{no system noise change} \\ > 1 : \text{system noise change} \end{cases}$$

The mathematical model of process in this case can be written in the form

$$\mathbf{x}(k+1) = \mathbf{f}(\mathbf{x}(k), k) + \mathbf{F}(k)\mathbf{w}(k), \tag{3.7}$$

where

$$\operatorname{diag}(\mathbf{F}(k)) = \begin{cases} (1 \quad 1 \quad . \quad . \quad . \quad 1), \quad \text{for } k < \tau \\ (\sigma_{11} \quad \sigma_{22} \quad . \quad . \quad . \quad \sigma_{ss}) \text{ if } \exists j \in \left(\overline{1, s}\right) \text{ where } \sigma_{jj} > 1 \text{ for } k \geq \tau \end{cases} \tag{3.8}$$

***Theorem 3.1*** If the measurements are processed by the EKF (A.1)–(A.7) and a process noise increment occurs at the iteration step $k = \tau$. Then at all $k \geq \tau$ steps, the process noise increment leads to an increment in the innovation covariance (A.4).

### 3.4.2 Influence of Process Noise Bias on the Innovation

***Assumption*** In this section, the process noise is assumed to be biased and can be represented as a of sum of random ($\mathbf{w}(k)$) and constant ($\boldsymbol{\delta}(k)$) components.

The process and measurement model equations in this case can be written as

$$\mathbf{x}_b(k+1) = \mathbf{f}(\mathbf{x}(k), k) + \boldsymbol{\delta}(k) + \mathbf{w}(k), \tag{3.9}$$

$$\begin{aligned} \mathbf{z}_b(k) &= \mathbf{H}\mathbf{x}_b(k) + \boldsymbol{\upsilon}(k) = \mathbf{H}[\mathbf{f}(\mathbf{x}(k), k) + \boldsymbol{\delta}(k) + \mathbf{w}(k)] + \boldsymbol{\upsilon}(k) \\ &= \mathbf{H}[\mathbf{f}(\mathbf{x}(k), k) + \mathbf{w}(k)] + \mathbf{H}\boldsymbol{\delta}(k) + \boldsymbol{\upsilon}(k) = \mathbf{H}\mathbf{x}(k) + \mathbf{H}\boldsymbol{\delta}(k) + \boldsymbol{\upsilon}(k) \\ &= \mathbf{z}(k) + \mathbf{H}\boldsymbol{\delta}(k) \end{aligned} \tag{3.10}$$

Here, b in the index shows that the value is biased.

It is clear that in this case the system state estimates will be biased.

***Theorem 3.2*** If the measurements are processed by the EKF (A.1)–(A.7) a process noise bias occurs at the iteration step $k = \tau$, then at all $k > \tau$ steps the estimation and innovation of EKF are biased and innovation bias is equal to the observed difference between the process bias and prediction bias.

A sampling covariance matrix of innovation is presented as statistics for detecting and compensating for the changes in process noise. The sample covariance matrix of the innovation may be expressed as follows (Hajiyev 2016):

$$\widehat{\mathbf{S}}_{\mathrm{e}}(k)=\frac{1}{M}\sum_{j=k-M+1}^{k}\widetilde{\mathbf{e}}(j)\widetilde{\mathbf{e}}^{T}(j) \tag{3.11}$$

where $M$ is the width of the "sliding window."

If there is a bias in the mean of the innovation at time $\tau$, and the biased innovation is denoted by $\widetilde{\mathbf{e}}_{\mathrm{b}}(k)$, then the biased innovation is defined as

$$\widetilde{\mathbf{e}}_{\mathrm{b}}(k)=\widetilde{\mathbf{e}}(k)\quad k=1,2,\ldots\tau-1 \tag{3.12}$$

$$\widetilde{\mathbf{e}}_{\mathrm{b}}(k)=\widetilde{\mathbf{e}}(k)+\boldsymbol{\mu}(k)\quad k=\tau,\tau+1,\ldots \tag{3.13}$$

where $\mu(.)$ is an unknown bias vector. When $k<\tau$, the mathematical expectation of the sample innovation covariance matrix (3.11) can be determined by the following equation:

$$E\left[\widehat{\mathbf{S}}_{\mathrm{e}}(k)\right]=\mathbf{P}_{\mathrm{e}}(k/k-1)=\mathbf{HP}(k/k-1)\mathbf{H}^{T}+\mathbf{R}(k) \tag{3.14}$$

In the case of $k\geq\tau$, in the sample innovation covariance (3.11) a biased innovation $\widetilde{\mathbf{e}}_{\mathrm{b}}(k)=\widetilde{\mathbf{e}}(k)+\boldsymbol{\mu}(k)$ is used instead of an unbiased innovation $\widetilde{\mathbf{e}}(k)$, where $\boldsymbol{\mu}(k)$ is the innovation bias (Hajiyev 2016)

$$\widehat{\mathbf{S}}_{\mathrm{eb}}(k)=\frac{1}{M}\sum_{j=k-M+1}^{k}\widetilde{\mathbf{e}}_{\mathrm{b}}(j)\widetilde{\mathbf{e}}_{\mathrm{b}}^{T}(j) \tag{3.15}$$

***Remark*** Note that the mean of the innovation $\widetilde{\mathbf{e}}_{\mathrm{b}}(k)$ in this case is not zero; therefore, formula (3.15) is not a sample covariance. In the "sliding window," it is the mean square of innovation (MSI).

***Statement*** The bias in innovation (A.3) leads to an increase in the mathematical expectation of the mean square of innovation.

Consequently, the process noise bias will increase the mathematical expectation of the mean square of innovation. It can be seen from *Theorem 3.2* and *Statement* above that the process noise bias is transferred to the innovation bias and changes the mean square of innovation (3.15). The innovation bias $\boldsymbol{\mu}(k)$ leads to a change in the values of $\widetilde{\mathbf{e}}_{\mathrm{b}}(k)$ and the elements of the mean square of innovation. As a result, the bias in the process noise is transferred to the MSI. Thus, the MSI can be chosen as a monitoring statistic in the problem of compensating for changes in process noise.

## 3.5 Analysis and Results

A tumbling small satellite is considered for the analysis in order to test the presented algorithms under crucial measurement and environmental challenges. The small satellite has the principal moment of inertia $J = \mathrm{diag}[0.055 \quad 0.055 \quad 0.017]$ kg m$^2$. The algorithm runs for almost 1 orbital period with 1 Hz sampling rate for the filter and the sensors. The sensors are selected as three-axis magnetometers and three-axis Sun sensors corrupted by the standard deviations of $\sigma_B = 300$ nT for magnetometers and $\sigma_S = 0.002$ for Sun sensors.

Only when at least two measurements are available at the same time and the vectors are not parallel to one another can the single-frame attitude estimate methods be used. As a result, when the satellite is shadowed and two vectors are parallel, the single-frame technique is ineffective. To demonstrate how the filter behaves that is integrated with a single-frame method in these intervals, an eclipse period is introduced between 2000th and 3000th seconds. This section evaluates the adaptation for process noise bias between 4500th and 5500th seconds by compring SVD-aided EKF (SaEKF) and adaptive SaEKF (ASaEKF). For investigating the filter's ability to adapt against process bias faults, the algorithm is tested. Process noise biases are identified by adding the constant bias term to the process noise between 4500th and 5500th seconds. For the test case, the constant term is selected as $\delta = [0.02 \quad 0.02 \quad 0.02 \quad 0.002 \quad 0.002 \quad 0.002]^T$

$$x(k+1) = f(x(k), k) + w(k) + \delta, \quad (4500 \leq k \leq 5500) \tag{3.16}$$

In Fig. 3.1, attitude angles estimations of ASaUKF and SaUKF algorithms are given when having noise biases. As seen, using the adaptive algorithm, the attitude angles can be estimated very well. The obtained results show that the bias-type process noise changes can be compensated using the covariance scaling techniques.

The root mean squares errors (RMS) for SaUKF and ASaUKF algorithms for process noise bias cases are shown in Table 3.1.

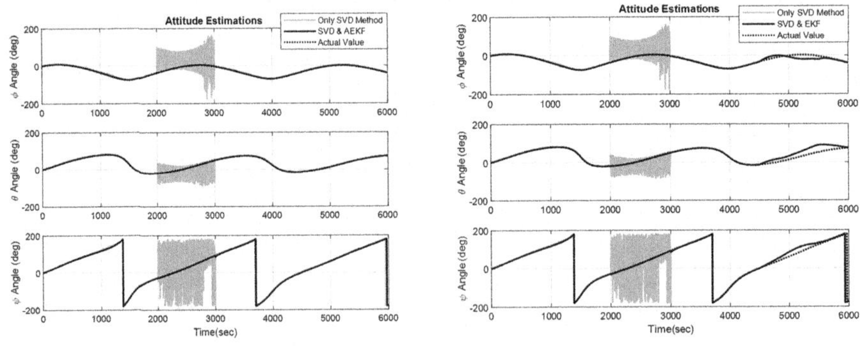


**Fig. 3.1** Attitude angles estimates of ASaUKF (**a**) and SaUKF (**b**) algorithms when applying the process noise bias

**Table 3.1** RMS error for SaUKF and ASaUKF for process noise bias case

| RMS error | Process noise bias $\delta$ | |
|---|---|---|
| | SaUKF | ASaUKF |
| $\phi$ | 14.6833 | 0.2897 |
| $\theta$ | 23.0296 | 0.2096 |
| $\psi$ | 18.5090 | 0.3199 |

## 3.6 Conclusions

In this chapter, an attitude filtering technique is proposed that adapts the process noise uncertainties. The bias and noise increment-type process noise uncertainties are taken into consideration. To begin with, the extended Kalman filter and singular value decomposition methods are combined to estimate a nanosatellite's attitude. The influence of the process noise bias and process noise increment-type system changes on the innovation of EKF is investigated. It is proved that the bias-type process noise change may be converted to the mean square of innovation of EKF and such types of changes can be compensatd using the covarince scaling techniques.

Various levels of process noise bias and process noise increment-type system changes are tested. Simulations are compared using the adaptive and nonadaptive versions of the nontraditional attitude filter. The simulation results show that in the investigated cases the multiple fading factors-based adaptive SVD-aided EKF can adapt to the changing environment better than the nonadaptive algorithm.

## Appendix A

### *EKF for Attitude and Rate Estimation*

In this Appendix, we restate the fundamental EKF equations applied to the attitude and rate estimation problem of nanosatellite. The state vector (3.5) estimate is

$$\widehat{\mathbf{x}}(k+1) = \widehat{\mathbf{x}}(k+1/k) + \mathbf{K}(k+1) \times \{\mathbf{z}(k) - \mathbf{H}\,\widehat{\mathbf{x}}(k+1/k)\} \tag{A.1}$$

where $\mathbf{z}(k) = \left[\mathbf{z}_{\Phi}(k) \quad \mathbf{z}_{\omega}(k)\right]^T$ is the measurement vector, and $\mathbf{H}$ is the measurement matrix. In our case, the measurement matrix is a $6 \times 6$ unit matrix. The extrapolation value is

$$\widehat{\mathbf{x}}(k+1/k) = \mathbf{f}[\widehat{\mathbf{x}}(k), k] \tag{A.2}$$

The innovation of EKF can be written as

$$\widetilde{\mathbf{e}}(k) = \mathbf{z}(k) - \mathbf{H}\,\widehat{\mathbf{x}}(k+1/k) = \mathbf{z}(k) - \mathbf{H}\,\mathbf{f}[\widehat{\mathbf{x}}(k), k] \tag{A.3}$$

The innovation covariance is

$$\mathbf{P}_{\mathrm{e}}(k+1/k) = \mathbf{H}\,\mathbf{P}(k+1/k)\mathbf{H}^{T} + \mathbf{R}(k) \tag{A.4}$$

The gain of the filter as a 6 × 6 matrix in size is

$$\mathbf{K}(k+1) = \mathbf{P}(k+1/k)\mathbf{H}^{T} \times \left[\mathbf{H}\,\mathbf{P}(k+1/k)\mathbf{H}^{T} + \mathbf{R}(k)\right]^{-1} \tag{A.5}$$

The covariance matrix of the extrapolation error as a 6 × 6 matrix in size is

$$\mathbf{P}(k+1/k) = \frac{\partial \mathbf{f}[\widehat{\mathbf{x}}(k), k]}{\partial \widehat{\mathbf{x}}(k)} \mathbf{P}(k/k) \frac{\partial \mathbf{f}^{T}[\widehat{\mathbf{x}}(k), k]}{\partial \widehat{\mathbf{x}}(k)} + \mathbf{Q}(k) \tag{A.6}$$

The covariance matrix of the filtering error as a 6 × 6 matrix in size is

$$\mathbf{P}(k+1/k+1) = [\mathbf{I} - \mathbf{K}(k+1)\mathbf{H}]\mathbf{P}(k+1/k) \tag{A.7}$$

# Appendix B

## *Q-Adaptive EKF*

The 6 × 1 innovation of the EKF is defined as in Eq. (B.1):

$$\widetilde{\mathbf{e}}(k) = \mathbf{z}(k) - \mathbf{H}\,\widehat{\mathbf{x}}(k+1/k) = \mathbf{z}(k) - \mathbf{H}\,\mathbf{f}[\widehat{\mathbf{x}}(k), k] \tag{B.1}$$

The innovation covariance at the iteration steps $k \geq \tau$ is

$$\begin{aligned} \mathbf{P}_{\mathrm{e}}(k+1/k) = & \mathbf{H}\left[\frac{\partial \mathbf{f}[\widehat{\mathbf{x}}(k), k]}{\partial \widehat{\mathbf{x}}(k)} \mathbf{P}(k/k) \frac{\partial \mathbf{f}^{T}[\widehat{\mathbf{x}}(k), k]}{\partial \widehat{\mathbf{x}}(k)}\right]\mathbf{H}^{T} \\ & + \mathbf{H}\mathbf{F}(k)\mathbf{Q}(k)\mathbf{F}^{T}(k)\mathbf{H}^{T} + \mathbf{R}(k) \end{aligned} \tag{B.2}$$

Since the matrices $\mathbf{F}(k)$ and $\mathbf{Q}(k)$ are assumed to be diagonal, the expression (B.2) can be rewritten in the following form:

$$\mathbf{P}_e(k+1/k) = \mathbf{H}\left[\frac{\partial\mathbf{f}[\widehat{\mathbf{x}}(k),k]}{\partial\widehat{\mathbf{x}}(k)}\mathbf{P}(k/k)\frac{\partial\mathbf{f}^T[\widehat{\mathbf{x}}(k),k]}{\partial\widehat{\mathbf{x}}(k)}\right]\mathbf{H}^T + \mathbf{H}\boldsymbol{\Lambda}(k)\mathbf{Q}(k)\mathbf{H}^T + \mathbf{R}(k) \tag{B.3}$$

where $\boldsymbol{\Lambda}(k) = \mathbf{F}^2(k)$ is the adaptive scaling matrix.

The scaling matrix $\boldsymbol{\Lambda}$ can be determined from equality (Hajiyev and Soken 2012):

$$\frac{1}{M}\sum_{j=k-M+1}^{k}\widetilde{\mathbf{e}}(j)\widetilde{\mathbf{e}}^T(j) \approx \mathbf{H}\,\mathbf{P}_e(k+1/k)\mathbf{H}^T + \mathbf{R}(k) \tag{B.4}$$

Here, $M$ is the width of the moving window. The left side of equality (B.4) represents the real, and the right side represents the theoretical value of the covariance of the innovation sequence. By substituting (B.3) into (B.4), we have

$$\begin{aligned}\frac{1}{M}\sum_{j=k-M+1}^{k}\widetilde{\mathbf{e}}(j)\widetilde{\mathbf{e}}^T(j) =& \mathbf{H}\left[\frac{\partial\mathbf{f}[\widehat{\mathbf{x}}(k),k]}{\partial\widehat{\mathbf{x}}(k)}\mathbf{P}(k/k)\frac{\partial\mathbf{f}^T[\widehat{\mathbf{x}}(k),k]}{\partial\widehat{\mathbf{x}}(k)}\right]\mathbf{H}^T \\ &+ \mathbf{H}\,\boldsymbol{\Lambda}(k)\,\mathbf{Q}(k)\,\mathbf{H}^T + \mathbf{R}(k)\end{aligned} \tag{B.5}$$

After performing the necessary mathematical transformations, the scale matrix $\boldsymbol{\Lambda}(k)$ can be determined from Eq. (B.5) as

$$\begin{aligned}\boldsymbol{\Lambda}(k) =& \left[\mathbf{H}^T\mathbf{H}\right]^{-1}\mathbf{H}^T \\ &\times\left\{\frac{1}{M}\sum_{j=k-M+1}^{k}\widetilde{\mathbf{e}}(j)\widetilde{\mathbf{e}}^T(j) - \mathbf{H}\left[\frac{\partial\mathbf{f}[\widehat{\mathbf{x}}(k),k]}{\partial\widehat{\mathbf{x}}(k)}\mathbf{P}(k/k)\frac{\partial\mathbf{f}^T[\widehat{\mathbf{x}}(k),k]}{\partial\widehat{\mathbf{x}}(k)}\right]\mathbf{H}^T - \mathbf{R}(k)\right\} \\ &\times\mathbf{H}\left[\mathbf{Q}(k)\mathbf{H}^T\,\mathbf{H}\right]^{-1}\end{aligned} \tag{B.6}$$

## References

Cilden D, Soken HE and Hajiyev C (2017) Nanosatellite attitude estimation from vector measurements using SVD-AIDED UKF algorithm. Metrology and Measurement Systems 24(1): 113–125. https://doi.org/10.1515/mms-2017-0011.

Hajiyev C (2016) An Innovation Approach Based Sensor Fault Detection and Isolation. IFAC-PapersOnLine 49(17). Elsevier: 420–425. https://doi.org/10.1016/J.IFACOL.2016.09.072.

Hajiyev C and Bahar M (2003) Attitude Determination and Control System Design of the ITU-UUBF LEO1 Satellite. Acta Astronautica 52(2–6): 493–499. https://doi.org/10.1016/S0094-5765(02)00192-3.

Hajiyev C and Cilden-Guler D (2017) Review on Gyroless Attitude Determination Methods for Small Satellites. Progress in Aerospace Sciences 90: 54–66. https://doi.org/10.1016/j.paerosci.2017.03.003.

Hajiyev C and Soken HE (2012) Robust Estimation of UAV Dynamics in the Presence of Measurement Faults. Journal of Aerospace Engineering 25(1): 80–89. https://doi.org/10.1061/(ASCE)AS.1943-5525.0000095.

Hajiyev C, Cilden D and Somov Y (2016) Gyro-free attitude and rate estimation for a small satellite using SVD and EKF. Aerospace Science and Technology 55: 324–331. https://doi.org/10.1016/j.ast.2016.06.004.

Ivanov D, Ovchinnikov M and Roldugin D (2018) Three-Axis Attitude Determination Using Magnetorquers. Journal of Guidance, Control, and Dynamics 41(11). American Institute of Aeronautics and Astronautics Inc.: 2455–2462. https://doi.org/10.2514/1.G003698.

Markley FL and Mortari D (2000) Quaternion Attitude Estimation using Vector Observations. Journal of the Astronautical Sciences 48(2): 359–380. https://doi.org/10.1007/BF03546284.

Mashtakov Y, Ovchinnikov M, Wöske F, et al. (2020) Attitude determination & control system design for gravity recovery missions like GRACE. Acta Astronautica 173. Elsevier Ltd: 172–182. https://doi.org/10.1016/j.actaastro.2020.04.019.

Mimasu BY and Van der Ha JC (2009) Attitude Determination Concept for QSAT. Transactions of the Japan Society for Aeronautical and Space Sciences, Aerospace Technology Japan 7: 63–68. https://doi.org/10.2322/tstj.7.Pd_63.

Springmann JC and Cutler JW (2014) Flight Results of a Low-cost Attitude Determination Systems. Acta Astronautica 99: 201–214. https://doi.org/10.1016/j.actaastro.2014.02.026.

# Chapter 4
# Exergetic Attribute of Fuel Cells in Hydrogen Aviation Including UAVs

**B. Kilkis**

## Nomenclature

| | |
|---|---|
| $\Delta E_X$ | Exergy destruction saving, kW-h |
| $\Delta W$ | Additional weight of onboard hydrogen system, kg |
| $\eta$ | First law efficiency |
| $E_X$ | Exergy, kW-h |
| GWP | Global warming potential |
| $H$ | Cruising height, m |
| $I_n$ | Solar irradiation on the panel, W/m$^2$ |
| ODI | Ozone depletion index |
| $Q$ | Thermal exergy, kW-h |
| TSI | Total solar irradiation, 1.36 W/m$^2$ |
| WV | Water vapor content, kg |

## 4.1 Introduction

Water vapor contributes to anthropogenic global warming, but as the global warming potential (GWP) is defined, it is assumed to be negligible for $H_2O$: an estimate gives a 100-year GWP between −0.001 and 0.0005, which is misleading.

$H_2O$ is the strongest greenhouse gas because it has a profound infrared absorption spectrum with more and broader absorption bands than $CO_2$. Its concentration in the atmosphere is limited by air temperature, so that radiative forcing by water vapor increases with global warming (positive feedback). But the GWP definition excludes indirect effects. The GWP definition is also based on emissions, and anthropogenic

B. Kilkis (✉)
OSTIM Technical University, Ankara, Turkiye



T. H. Karakoc et al. (eds.), *The Future of Electric Aviation and Artificial Intelligence*, Sustainable Aviation, https://doi.org/10.1007/978-3-031-62094-2_4

emissions of water vapor are removed via precipitation within weeks, so its GWP is negligible. We need another definition: ozone depletion index (ODI) (Kilkis 2019a). Therefore, in UAV applications extra care must be taken to release only oxygen into the atmosphere. Nitrogen ($N_2$) and oxygen ($O_2$) molecules do not absorb infrared photons even though they make up more than 90% of Earth's atmosphere.

## 4.2 Exergetic Model

The system must be optimized for the added value of extra electrolysis/hydrogen/TEG systems and exergy destruction avoidances versus penalty of extra weight. For each climb to an altitude of $H$, the extra system weight $\Delta W$ needs an exergy of $0.95\ W \times H$. Avoidance of exergy destructions with new hydrogen systems, $\Delta E_X$, during a single flight must obey the following constraints:

$$|\Delta E_X| > 0.95 \times \Delta W \times H \quad \{\text{Exergy Balance}\} \tag{4.1}$$

$$-(2.1 \times \Delta E_X + aWV^n) > 0.23 \times (0.95 \times \Delta W \times H) \quad \{\text{Emission Balance}\} \tag{4.2}$$

In Eq. (4.1), the first term represents the exergy destruction avoidance on board by the additional equipment as shown in Fig. 4.1. The second term on the right-hand side of Eq. (4.1) represents the exergy spent during the climb. In Eq. (4.2), the left-hand side has two terms. The first term is concerned with the exergy destruction savings with a multiplier of 2.1, which is the multiplier for the exergy destruction-to-$DCO_2$ term when the destruction takes place upstream of the process involved. The second term represents the $CO_2$ equivalence of water vapor captured before releasing into the atmosphere. Figure 4.1 shows the background of green hydrogen use by the

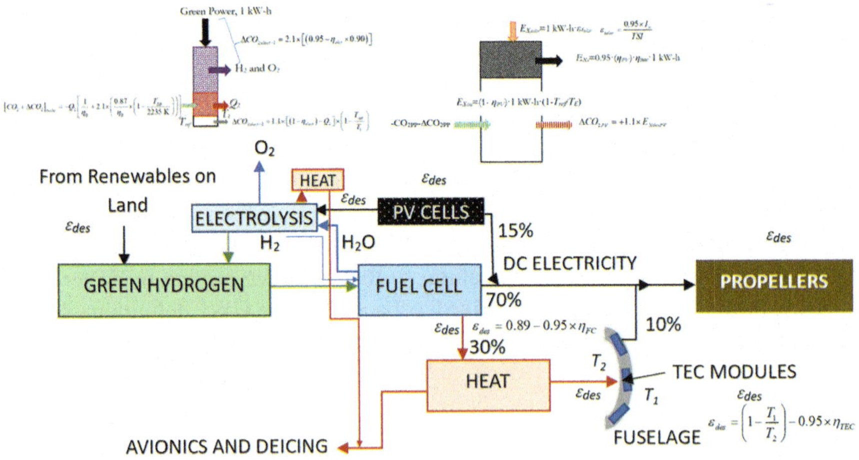


**Fig. 4.1** Air side: hydrogen use in a UAV with only $O_2$ release

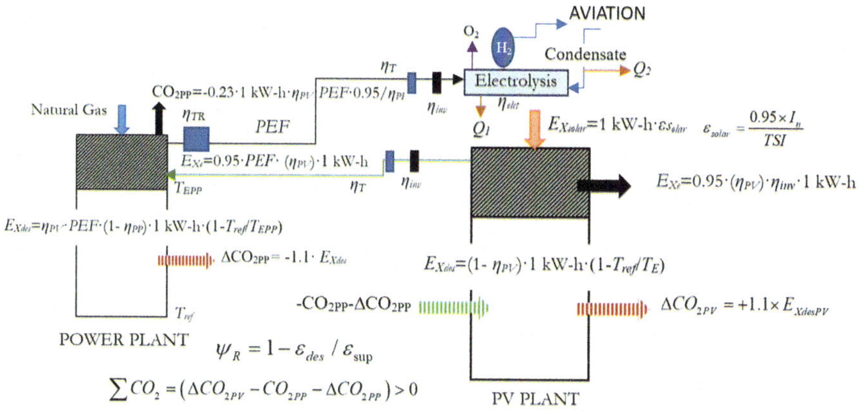


**Fig. 4.2** History of green hydrogen on the land side

UAV on the air side during takeoff, cruise, and landing. DW also affects the power demand in level flight (Kilkis et al. 2019).

The history of green hydrogen on the land side is shown in Fig. 4.2. Hydrogen may also be derived from $H_2S$ gas and other pollutants (Kilkis 2019b).

## 4.3 Conclusion

- Green hydrogen fuel in UAVs has certain advantages.
- The water vapor release must be avoided or minimized for environmental reasons both on the ground side and the air side.
- In this effort, the second law of thermodynamics must be employed.

## References

Kilkis, B., Kilkis, San., and Kilkis Siir. (2019) A Simplistic Flight Model for Exergy Embodiment of Composite Materials Towards Nearly-Zero Exergy Aviation, International Journal of Sustainable Aviation 5(1):19. https://doi.org/10.1504/IJSA.2019.099922.

Kilkis, B. (2019a) A Model for Combining the Ozone Depletion Potential of Refrigerants, 3rd National Polar Science Workshop, 5–6 November, METU Ankara.

Kilkis, B. (2019b) A Holistic View of Sustainable Aviation, Chapter 5 in: *Sustainable Aviation*, (Eds: Karakoç, Çolpan, Altuntaş, Şöhret), pp: 97–121, Springer Nature, Switzerland, ISBN 978-3-030-14194-3 ISBN 978-3-030-14195-0 (eBook), https://doi.org/10.1007/978-3-030-14195-0

# Chapter 5
# Direct Numerical Simulation of a Thermal Turbulent Boundary Layer: An Analogy to Simulate Bushfires and a Testbed for Artificial Intelligence Remote Sensing of Bushfire Propagation

**Julio Soria, Shahram Karami, Callum Atkinson, and Minghang Li**

## Nomenclature

| | |
|---|---|
| AI | Artificial intelligence |
| CNN | Convolutional neural networks |
| DL | Deep learning |
| DNS | Direct numerical simulation |
| FUEGO | Fire urgency estimator in geosynchronous orbit |
| PiDL | Physics-informed deep learning |
| RNN | Recurrent neural networks |
| SciML | Scientific machine learning |
| TTBL | Thermal turbulent boundary layer |
| UAV | Unmanned aerial vehicle |
| UQ | Uncertainty quantification |
| ZPG | Zero-pressure-gradient |

## 5.1 Introduction

Australia, like many other places in the world with a similar climate, experiences frequent bushfires or wildfires, which, as was recently experienced, pose a serious threat to the local population, unique Australian wildlife, and natural resources, as

J. Soria (✉) · S. Karami · C. Atkinson · M. Li
Laboratory for Turbulence Research in Aerospace and Combustion, Department of Mechanical and Aerospace Engineering, Monash University, Clayton Campus, Melbourne, VIC, Australia
e-mail: julio.soria@monash.edu; shahram.karami@monash.edu; callum.atkinson@monash.edu; minghang.li@monash.edu



T. H. Karakoc et al. (eds.), *The Future of Electric Aviation and Artificial Intelligence*, Sustainable Aviation, https://doi.org/10.1007/978-3-031-62094-2_5

**Fig. 5.1** Blue Mountain bushfire (Gospers Mountain, NSW Australia, December 2019) burned more than 510,000 ha (1.26 million acres), making it the largest forest fire ever recorded in Australia (Koumoundouros 2022; Bush Fire Bulletin 2020)

illustrated in Fig. 5.1 (Koumoundouros 2022; Damany-Pearce et al. 2022), in addition to contributing to an increase in global $CO_2$ emissions.

Despite extensive research, there is still no definitive theory to accurately predict the dynamics and spread of bushfires. This gap in understanding impedes the development of effective prevention, mitigation, and control strategies. Existing computational models and numerical weather predictions, cited as inadequate by Binskin (2020) and Mayfield (2020), fall short for real-time operational use.

Current remote monitoring methods rely on infrared satellite sensing, which provide temperature data but do not offer insights into the fire's rate and direction of movement within the complex atmospheric conditions. To improve disaster management, we need to supplement this temperature data with predictive capabilities for fire behavior and spread.

This chapter introduces a specialized scientific machine learning (SciML) approach (Baker et al. 2019; Willcox 2019; Fukami et al. 2021) that incorporates physics-informed machine learning based on deep learning (DL). This methodology aims to assimilate remotely sensed temperature data to provide augmented information on key bushfire dynamics, like wind velocity and energy transport. The real-time availability of such enriched data could revolutionize all phases of bushfire disaster management by offering high-accuracy, low-uncertainty predictions currently unavailable.

Validating this SciML approach requires high-quality data, ideally from direct numerical simulations (DNS) of turbulent bushfire behavior. However, performing DNS in real-world scenarios is computationally unfeasible, even with advanced supercomputers, given the complexity and large parameter space. As a workaround, this chapter proposes using a simulated “ground truth” model to serve as a testbed for our AI-enhanced remote sensing methodology.

The chapter outlines the SciML methodology and its uncertainty quantification (UQ), describes an alternative DNS methodology for “ground truth” validation by using a complex multi-scale, highly dynamic, thermal turbulent boundary layer (TTBL) DNS of the atmosphere that considers various energy sources and temperature-dependent heat-release models (Li et al. 2020), and presents one such simulated example for testing AI-enhanced remote sensing approaches to bushfire monitoring.

## 5.2 SciML Based on Physics-Informed Deep Learning (PiDL) for Bushfire Predictions

There is significant effort into remote sensing for the detection and prediction of bushfires/wildfires, such as the Fire Urgency Estimator in Geosynchronous Orbit (FUEGO) Pennypacker et al. (2013). However, there is currently no operational nor proposed system that employs AI in the form of discipline-specific SciML to enhance this data and provide bushfire dynamics and reliable fire-front propagation predictions. The proposed discipline-specific SciML will employ remote sensing by satellites, fixed aircraft, UAV, etc., fitted with infrared sensors to acquire 3D temperature field information of bushfires as outlined by Pennypacker et al. (2013).

Figure 5.2a shows the proposed SciML based on physics-informed DL (PiDL) for bushfire predictions. The inputs to the PiDL are the four-dimensional vector space of three-dimensional (3D) space and time $(x, y, z, t)$, which can be defined on an arbitrary three-dimensional unstructured grid, and the sampled 3D temperature data $T_m(x_m, y_m, z_m, t_m)$, which can come from a range of sources. The PiDL uses this data to both post-dict (via training) and predict the TTBL turbulent velocity vector field, the turbulent pressure and temperature fields, as well as the source term, which represents the heat release rate due to burning of the biomass during the bushfire.

Within the PiDL is the DL subsystem, which can be a fully connected neural network, convolutional neural networks (CNNs), or recurrent neural networks (RNNs), etc. The disciple-specific physics-information is in the form of the incompressible Navier–Stokes equations and the temperature energy equation coupled via a Boussinesq approximation.

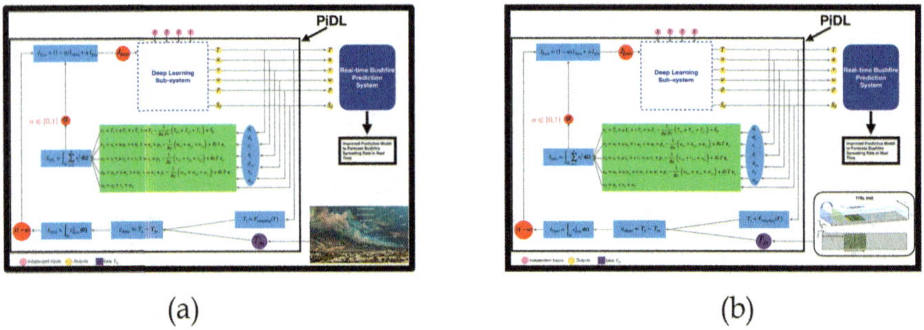


**Fig. 5.2** (**a**) Proposed SciML based on PiDL for bushfire predictions. The components of the system within the black rectangular box represent the PiDL. The inputs to the PiDL are the independent four-dimensional vector space of space and time $(x, y, z, t)$ and the sampled 3D temperature data $T_m$, which is here pictorially represented from data that is obtainable from the FUEGO vertical envelope (Pennypacker et al. 2013). The AI-learned output of the PiDL is the TTBL turbulent velocity vector field, the 3D turbulent temperature field, and the source term that represents the heat release due to burning of the biomass during the bushfire. All outputs are provided by the PiDL as functions of four-dimensional vector space of space and time $(x, y, z, t)$. (**b**) Development, testing, and UQ of the proposed SciML based on PiDL for bushfire predictions using "ground truth" TTBL DNS temperature data

### *5.2.1 "Ground Truth" Data and Uncertainty Quantification of PiDL*

The detailed development of the PiDL relies on "*ground truth*" data. However, this type of data for a bushfire is impossible to acquire from real bushfires or even from limited experimental laboratory "bush" fires. However, DNS of a TTBL with distributed arbitrary energy sources with a well-defined temperature-dependent heat-release rate model as a source term in the energy equation (Li et al. 2020) can serve as an analog to a bushfire, providing the full highly resolved turbulent velocity vector field, the pressure and temperature field, and the heat release source term as the necessary "*ground truth*" data as shown in Fig. 5.2b, with the approach to UQ of the PiDL illustrated in Fig. 5.3.

## 5.3 Direct Numerical Simulation of a Thermal Turbulent Boundary Layer: An Analogy to Simulate Bushfires

DNS is used to accurately simulate a canonical TTBL case, which is an analogy to a bushfire as a first step by including a fully resolved zero-pressure-gradient (ZPG) TBL with the bushfire modeled as a distribution of localized heat sources with a temperature-dependent energy release rate (Li et al. 2020).

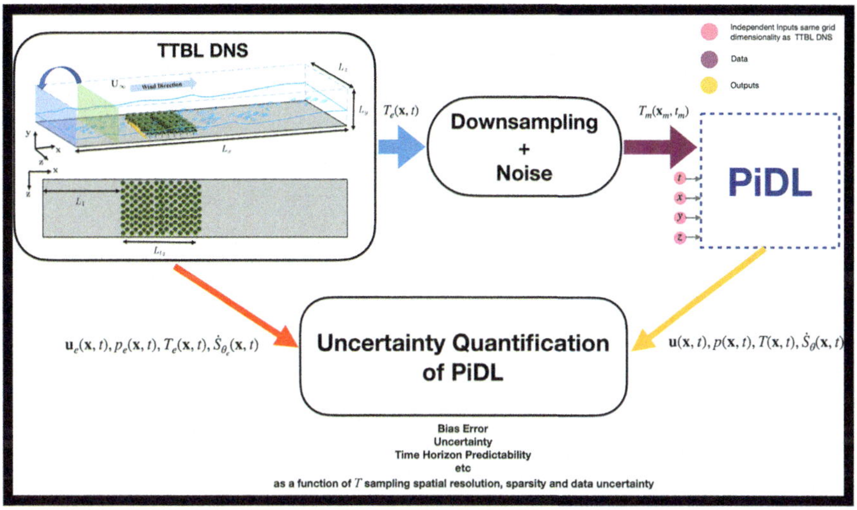


**Fig. 5.3** System diagram for UQ of the proposed SciML based on PiDL for bushfire predictions using "ground truth" TTBL DNS data. Variables with the subscript "$e$" indicate TTBL DNS data, variables with the subscript "$m$" indicate simulated measured temperature data, while variables with no subscript indicate output variables from the PiDL

### *5.3.1 Theoretical Framework of the TTBL DNS*

The governing equations for the TTBL are the incompressible Navier–Stokes equations with the temperature–energy equation including an energy source term to model the heat release rate by the burning of the biomass representing the bush, etc. The Navier–Stokes equations are coupled to the temperature–energy equation via a Boussinesq approximation (Tritton 1977) and are given by

$$\frac{\partial u_i}{\partial x_i} = 0 \tag{5.1}$$

$$\frac{\partial u_i}{\partial t} + \frac{\partial u_i u_j}{\partial x_j} = -\frac{\partial p}{\partial x_i} + \frac{1}{\text{Re}}\frac{\partial^2 u_k}{\partial x_k^2} + \frac{\kappa_i}{\text{Fr}_\text{o}{}^2}\theta \tag{5.2}$$

$$\frac{\partial \theta}{\partial t} + \frac{\partial u_j \theta}{\partial x_j} = \frac{1}{\text{RePr}}\frac{\partial^2 \theta}{\partial x_i^2} + \dot{S}_\theta, \tag{5.3}$$

The last term on the right-hand side of Eq. (5.2) is the Boussinesq approximation, with $\kappa_i$ a vector indicating the direction of the gravitational field, which in this case is $[0, -1, 0]$, while the last term on the right-hand side of Eq. (5.3) represents the source term, which represents the heat release rate from the burning of biomass sources. This term needs to be modeled. Equations (5.1)–(5.3) are written in nondimensional form with respect to the nondimensional spatial vector $x_j$, velocity vector $u_i$, pressure

$p$, and temperature $\theta$, defined with respect to its dimensional starred independent and dependent variables as

$$x_j = \frac{x_j^*}{L}; \quad u_j = \frac{u_j^*}{U_\infty^*}; \quad \rho = \frac{\rho^*}{\rho_\infty^*}; \quad t = \frac{t^* U_\infty^*}{L}; \quad p = \frac{p^*}{\rho_\infty^* U_\infty^{*2}}; \quad \theta = \frac{T^* - T_\infty^*}{T_{\mathrm{ad}}^* - T_\infty^*}$$
$$\zeta_0 = \frac{T^*}{T_{\mathrm{ad}}^* - T_\infty^*}; \quad \mu = \frac{\mu^*}{\mu_\infty^*}; \mathrm{Pr} = \frac{\nu_\infty^*}{\alpha_\infty^*}; \quad \mathrm{Fr} = \frac{U_\infty^*}{\sqrt{gL}}; \quad \mathrm{Re} = \frac{\rho_\infty^* U_\infty^* L}{\mu_\infty^*};$$
$$\mathrm{Fr}_0 = \zeta_0^{1/2} \mathrm{Fr} \tag{5.4}$$

where the characteristic quantities are: $L$ = characteristic length, $U_\infty^*$ = characteristic free-stream velocity, $\rho_\infty^*$ = characteristic density, $T_\infty^*$ = characteristic free-stream temperature, $\mu_\infty^*$ = characteristic dynamic viscosity, $\nu_\infty^* = \frac{\mu_\infty^*}{\rho_\infty^*}$ = characteristic kinematic viscosity, $\alpha_\infty^*$ = characteristic thermal diffusivity, $g$ = acceleration due to gravity, and $T_{\mathrm{ad}}^*$ = characteristic adiabatic flame temperature. The nondimensional parameters are Re = Reynolds number, Pr = Prandtl number, and Fr = Froude number.

The source term $\dot{S}_\theta$ in Eq. (5.3), which accounts for the heat release rate of the burning of biomass, is for simplicity modeled as an exothermic reaction of the type $F \longrightarrow P$, where $F$ is the unburned biomass and $P$ is the product. Furthermore, it is assumed that the biomass and product have the same constant heat capacity, molecular weight, and molecular diffusion coefficient. It can be shown that the source term can be written as

$$\dot{S}_\theta = A(1-\theta) \exp\left[\frac{-\beta(1-\theta)}{1-\alpha(1-\theta)}\right], \tag{5.5}$$

where $A$ is a constant. This approach is the center of many early developments in theoretical combustion (Clavin and Williams 1979; Bychkov and Liberman 2000). Despite its strong assumptions, especially at the chemistry level, this source term model preserves many features, including nonlinear heat release and variable temperature.

### 5.3.2 *Numerical Method*

The high-fidelity simulation of TTBL is conducted using a modified version of a hybrid parallel MPI/OpenMP TBL DNS code (Simens et al. 2009; Borrel et al. 2013; Kitsios et al. 2016; Kitsios et al. 2017). The code solves the three-dimensional incompressible Navier–Stokes equations in a three-dimensional rectangular volume. The bottom surface is a flat plate with a no-slip boundary condition. The thermal boundary condition can be specified as a constant temperature or a constant heat flux. The former is used in the TTBL DNS results presented here.

### 5.3.3 Results

Figure 5.4 shows the full computational domain of TTBL and a zoomed-in version of the biomass domain. Figure 5.5 shows three snapshots along the bushfire simulation, beginning from the fire starting, evolving as a bushfire without too much of a

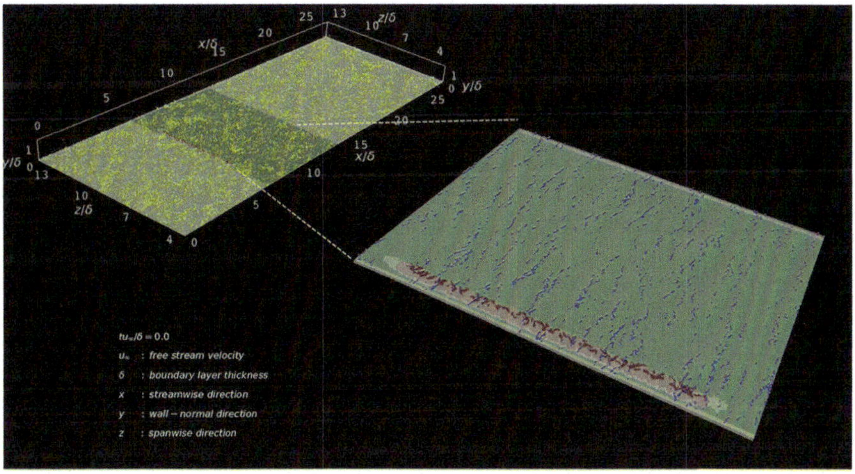


**Fig. 5.4** Visualization of the TTBL DNS, with the green region indicating the biomass fuel, the yellow structures in the full computational domain being coherent vortical structures visualized using the second invariant of the velocity gradient tensor (Ooi et al. 1999), while the blue streaks in the zoomed-in view represent low-velocity streaks. The red color in both indicates the bushfire front at the beginning of the bushfire. A full animation can be viewed in Li et al. (2020)

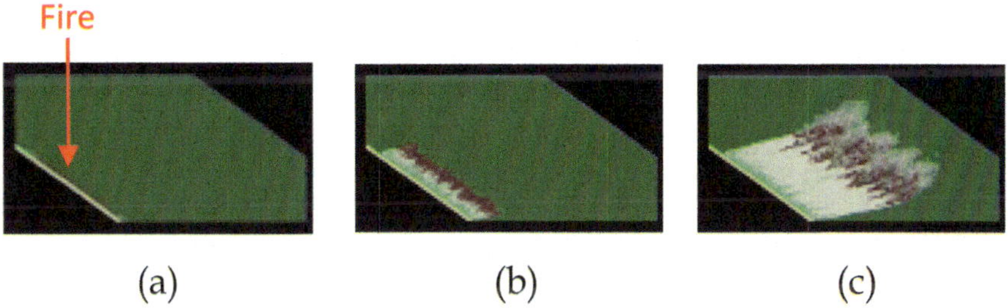


**Fig. 5.5** Snapshots of the biomass burning evolution zoomed-in from Fig. 5.4. (**a**) indicates the starting of the bushfire, (**b**) is a time after the bushfire has started, but with little smoke at this stage, and (**c**) indicates the progress of the bushfire, with the gray color indicating fluid temperatures that are typical of smoke rather than fire

smoke bloom, and subsequently evolving into a bushfire, with significant smoke reaching the upper parts of the TTBL. A full animation of the bushfire analogy provided by this TTBL DNS can be viewed in Li et al. (2020). It is worth noting the qualitative similarities in the latter stages of the TTBL DNS and the photograph of the Blue Mountain bushfire given in Fig. 5.1.

## 5.4 Concluding Remarks

This chapter proposes a SciML methodology that utilizes physics-informed machine learning based on deep learning (DL)-PiDL to assimilate remotely sensed temperature data and the approach to its UQ. The development, testing, and UQ are entirely reliant on the TTBL DNS, which is a canonical analogy of a bushfire that provides training data and the "*ground truth*" for UQ. Ultimately in the field, the PiDL will learn from remotely sensed 3D temperature data gathered from a variety of sources, ranging from satellites to fixed-wing aircraft and UAVs or drones fitted with infrared sensors and be able to predict the thermal turbulent boundary layer associated with a bushfire, as well as predict its dynamics and fire-front spreading rate. A realistic heat source rate model has been developed based on approaches that have their roots in the classical developments of theoretical combustion. This model has been employed to undertake an initial canonical bushfire analogy using a uniform biomass distribution. The resulting TTBL DNS has shown close qualitative aspects with images of the recent Blue Mountain bushfire in Australia.

**Acknowledgments** This research was supported by an NCI Australasian Leadership Computing Grant and a Monash Data Futures Institute Seed Grant. The authors also gratefully acknowledge the NCMAS HPC allocation of NCI supported by the Australian federal government and of Pawsey, also supported by the Australian federal government and the state government of Western Australia.

## References

Baker, N et al. (2019) Tech. rep. United States: US Department of Energy, Office of Science (SC).
Binskin, M (2020) Royal Commission into Natural Disaster Arrangements, 594 pages, https://nla.gov.au/nla.catbibrangedashvn8555232
Borrel, G., Sillero, JA and Jimenez, J (2013) A code for direct numerical simulation of turbulent boundary layers at high Reynolds numbers in BG/P supercomputers. Comp. Fluids, 80:37–43.
Bychkov, VV and Liberman, MA (2000) Dynamics and stability of premixed flames. Physics reports, 325(4–5):115–237.
Bush Fire Bulletin (2020) Sydney: NSW Rural Fire Service 42.1, http://www.rfs.nsw.gov.au/__data/assets/pdf_file/0007/174823/Bush-Fire-Bulletin-Vol-42-No1.pdf
Clavin, P and Williams, FA (1979) Theory of premixed-flame propagation in large-scale turbulence. Journal of Fluid Mechanics, 90(3):589–604.
Damany-Pearce, L, Johnson, B and Wells, A. (2022) Sci Rep 12, p. 12665

Dold, JW, Thatcher, RW, Omon-Arancibia, A, and Redman, J (2002) From one-step to chain-branching premixed flame asymptotics. Proceedings of the Combustion Institute, 29(2): 1519–1526.

Fukami, K., Fukagata, K., & Taira, K. (2021). Machine-learning-based spatio-temporal super resolution reconstruction of turbulent flows. Journal of Fluid Mechanics, 909, A9. https://doi.org/10.1017/jfm.2020.948

Kitsios, V, Sekimoto, A, Atkinson, C, Sillero, JA, Borrell, G, Gungor, AG, Jiménez, J, and Soria, J (2017) Direct numerical simulation of a self-similar adverse pressure gradient turbulent boundary layer at the verge of separation. J. Fluid Mech., 829:392–419.

Kitsios, V, Atkinson, C, Sillero, JA, Borrell, G, Gungor, AG, Jimenez, J and Soria, J (2016) Direct numerical simulation of a self-similar adverse pressure gradient turbulent boundary layer. Int. J. Heat Fluid Flow, 61(A):129–136.

Koumoundouros, T (2022) https://www.sciencealert.com/australias-massive-bushfires-spawned-a-dramatic-heat-anomaly-in-the-stratosphere

Li, M et al. (2020) https://doi.org/10.1103/APS.DFD.2022.GFM.V0090

Mayfield, P (2020) CSIRO Climate and Disaster Resilience: Technical Report, https://www.csiro.au/-/media/Environment/CSIRO-Climate-and-Disaster-Resilience-Technical.pdf

Ooi, A, Martin, J, Soria, J and Chong, MS (1999) A study of the evolution and characteristics of the invariants of the velocity-gradient tensor in isotropic turbulence. J. Fluid Mech. 381, 141–174.

Pennypacker, C, Jakubowski, M, Kelly, M, Lampton, M, Schmidt, C, Stephens, S and Tripp, R (2013) FUEGO — Fire Urgency Estimator in Geosynchronous Orbit — A Proposed Early-Warning Fire Detection System. Remote Sensing 5 5173–92, https://doi.org/10.3390/rs5105173

Peters, N (1997) Kinetic foundation of thermal flame theory. Advances in combustion science: In honor of Ya. B. Zel'dovich (A 97-24531 05-25), Reston, VA, American Institute of Aeronautics and Astronautics, Inc. (Progress in Astronautics and Aeronautics., 173:73–91.

Simens, MP, Jimenez, J, Hoyas, S, and Mizuno, Y (2009) A high-resolution code for turbulent boundary layers. J. Comp. Phys., 228:4128–4231.

Tritton, DJ (1977). Physical fluid dynamics. New York: Van Nostrand Reinhold Co.

Willcox K (2019) In: International Congress on Industrial and Applied Mathematics ICIAM2019, July 15–19, Valencia, Spain.

# Chapter 6
# Unmanned Aerial Vehicle Cooperation for the Monitoring of Greenhouse Gases

**Glauco Caurin, Antonio Carlos Daud Filho, and José Reinaldo Silva**

## Nomenclature

| | |
|---|---|
| ATTO | Amazon Tall Tower Observatory |
| GHG | Greenhouse gas |
| INPA | National Institute for Research in the Amazon |
| INPE | National Institute for Space Research |
| MCTIC | Brazilian Ministry of Science, Technology, Innovation and Communications |
| UAV | Unmanned aerial vehicle |
| UEA | State University of Amazonas |

## 6.1 Introduction

The Amazon Forest region is fundamental for the global balance of greenhouse gases (GHG) since it is the largest continental carbon reservoir. In this way, the deforestation of tropical forests, such as the Amazon rainforest (Sierra et al. 2023), can have a considerable impact on carbon emissions. In this context, the deforestation of tropical forests is responsible for about 13% of global $CO_2$ emissions. What's more, deforestation in Brazil has increased sharply over the last 5 years.

G. Caurin (✉)
University of São Paulo, São Carlos School of Engineering, Department of Aeronautical Engineering, São Carlos, Brazil
e-mail: gcaurin@sc.usp.br

A. C. Daud Filho · J. R. Silva
University of São Paulo, Polytechnic School, Department of Mechatronics and Mechanical Systems Engineering, São Paulo, Brazil
e-mail: antonio.daud.filho@usp.br; reinaldo@usp.br



T. H. Karakoc et al. (eds.), *The Future of Electric Aviation and Artificial Intelligence*, Sustainable Aviation, https://doi.org/10.1007/978-3-031-62094-2_6

**Fig. 6.1** Three views of the Amazon Tall Tower Observatory (ATTO 2023)

Thus, there is a need to better understand the sources and sinks of carbon cycling in the Amazon rainforest region, as well as the reforestation of this region can have enormous potential for global carbon absorption. This region encompasses a huge area not only in Brazil, but also in other countries in South America.

There are currently relevant data acquisition initiatives associated with GHG in the Amazon rainforest region. One example, implemented on the surface of the forest, is the Amazon Tall Tower Observatory (ATTO) (Andreae et al. 2015; ATTO 2023; LBA 2023), a joint venture of INPA/MCTIC (National Institute for Research in the Amazon/Brazilian Ministry of Science, Technology, Innovation and Communications), in partnership with the Max Planck Institute in Germany, the State University of Amazonas (UEA), among other institutions. An image of the ATTO is shown in Fig. 6.1. But most of the data collected come from satellites. The National Institute for Space Research (INPE) extensively uses data from satellites such as Landsat to provide data to monitor the huge Amazon area. Experiences with these initiatives have generated good results and a better understanding of the challenges ahead. This motivated the scientific community's demand for new missions and improvement.

Given this current state, we are looking for new possibilities in order to increase the availability of data and the frequency with which some areas are monitored, especially when we think in terms of data processing information update frequency and response latency. There is also, for example, the possibility of offering measurements of GHG that go beyond the average values of each area and start offering measurements at different altitudes. This measurement system at various altitudes, using different vehicles, is called the UXV Ecosystem.

## 6.2 UXV Ecosystem

In this context, in order to obtain comprehensive and reliable GHG data in the most diverse situations present in the Amazon Forest region, it is necessary to have a wide range of tools and vehicles with complementary performance capabilities and data collection availability. Figure 6.2 shows the concept of the UXV Ecosystem, which makes use of several types of vehicles depending on the altitude of the data collection operation.

Thus, the application of the UXV Ecosystem will enable missions to be carried out in order to monitor the situation of the forest. The missions, as well as the proposed platforms, are listed below.

- *Mission:* precise assessment of deforestation.

  *Proposed platform:* integrated operation of an unmanned airship and long-range autonomous aircraft.
- *Mission:* accurate assessment of environmental damage due to fire and early warning of fire.

  *Proposed platform:* combination of unmanned airship, a swarm of UAVs and a high-altitude UAV.
- *Mission:* collection of river water samples and acquisition of greenhouse gas samples in close proximity to the water surface.

  *Proposed platform:* combined operation of boat- or ship-based station together with an ekranoplane-type aircraft.
- *Mission:* continuous measurement of GHG concentrations up to 15 km.

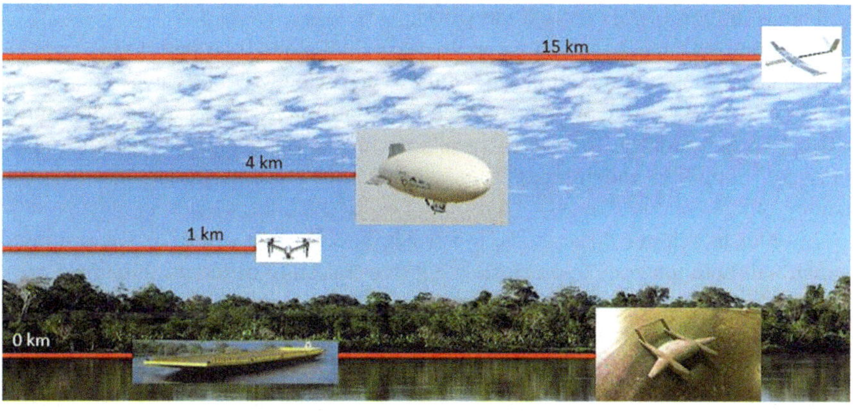


**Fig. 6.2** Illustrative concept of the multiplatform system of systems proposed to monitor the forest

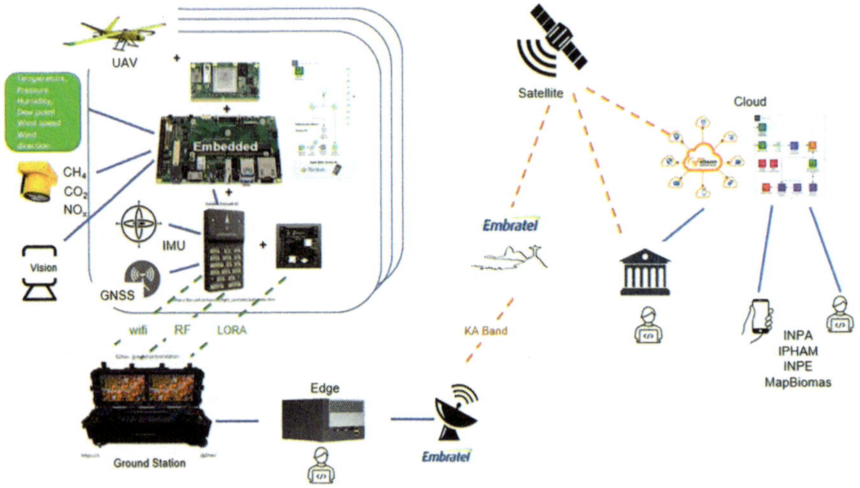


**Fig. 6.3** Architecture of the proposed GHG data acquisition system

*Proposed platform:* high-altitude UAV.

Thus, to improve assessment procedures and their results goes beyond current data acquisition systems, paradigms, and platforms, that is, stationary towers, satellites, or isolated instrumented flights are not sufficient to monitor the forest. As one may observe from the previous proposed missions, it is not a trivial task, and it is definitely not a task for a single platform or aircraft.

This entire ecosystem of platforms in space, air, land, and water can produce a significant amount of data. Thus, we imagine that data from preexisting towers and satellites will receive additional contributions from the unmanned aerial vehicle ecosystem. An important challenge is to extract data from the forest and take them to the computers of researchers in the environmental, cartographic, and meteorological areas in their offices around the world.

Communication between unmanned aerial vehicles and base stations incorporates new technologies and is continuously improving, which is good news, considering that new sensors need to be incorporated into platforms, such as sensors measuring $CO_2$, $CH_4$, in addition to conventional meteorological quantities of temperature, pressure, humidity, and images from conventional cameras, Lidar, multispectral, among others.

In order to achieve these goals, the system architecture of Fig. 6.3 is proposed, which will help integrate and coordinate the operation of different unmanned aerial vehicles collecting data, in addition to testing ways to extract data collected from the forest in a realistic and viable way, take them to our computers at the university, and share them with other research centers using cloud data processing infrastructure.

Thus, the current research is focused on the development of an integrated GHG data acquisition system that can be embarked on an unmanned vehicle, and this must be able to cover a trajectory in the Amazonian Forest region carrying out the measurements. The acquired data must be shared through a cloud computational architecture. Thus, the research must determine which tasks of processing the acquired data must be performed on board the vehicle and which must be performed by the end user of the data.

Therefore, the development of trajectory planning algorithms for autonomous aerial exploration is also part of the current research. In an online environment, the path to be followed must optimize the search by observing quantities of interest such as computed trails, temperature, pollution, and deforestation in order to find the best route, the quality of which is determined by the amount of space, and uncharted area that can be exploited, taking into account the range constraints of the aircraft. The proposed planner must be able to be processed online, on board the vehicle, with limited resources. This high performance should be evaluated in detailed simulation studies as well as in real-world experiments using an unmanned aerial vehicle. Analysis of the algorithm's computational complexity should also be done to verify the proposal's ability to deal with large-scale issues.

## 6.3 Conclusion

Therefore, in order to obtain comprehensive and reliable GHG data in the most diverse situations present in the Amazon Forest region, it is necessary to have a wide range of tools and vehicles with complementary performance capabilities and data collection availability, so that it is possible to develop consistent models and analyze and assess the situation of the forest. In this sense, this project seeks to provide the tools and structure that enable these objectives to be achieved.

**Acknowledgments** The authors acknowledge the AWS and FAPESP grants 2018/15472-9, 2013/07276-1, and 2020/15230-5. This study was also partially funded by the CAPES and FINEP in Brazil.

## References

Andreae, M. O., Acevedo, O. C., Araùjo, A., Artaxo, P., Barbosa, C. G. G., Barbosa, H. M. J., Brito, J., Carbone, S., Chi, X., Cintra, B. B. L., da Silva, N. F., Dias, N. L., Dias-Júnior, C. Q., Ditas, F., Ditz, R., Godoi, A. F. L., Godoi, R. H. M., Heimann, M., Hoffmann, T., Kesselmeier, J., Könemann, T., Krüger, M. L., Lavric, J. V., Manzi, A. O., Lopes, A. P., Martins, D. L., Mikhailov, E. F., Moran-Zuloaga, D., Nelson, B. W., Nölscher, A. C., Santos Nogueira, D., Piedade, M. T. F., Pöhlker, C., Pöschl, U., Quesada, C. A., Rizzo, L. V., Ro, C.-U., Ruckteschler, N., Sá, L. D. A., de Oliveira Sá, M., Sales, C. B., dos Santos, R. M. N., Saturno, J., Schöngart, J., Sörgel, M., de Souza, C. M., de Souza, R. A. F., Su, H., Targhetta, N., Tóta, J., Trebs, I., Trumbore, S., van Eijck, A., Walter, D., Wang, Z., Weber, B., Williams, J., Winderlich, J., Wittmann, F., Wolff, S., and Yáñez-Serrano, A. M. (2015) The Amazon Tall Tower Observatory (ATTO): overview of pilot measurements on ecosystem ecology, meteorology, trace gases, and aerosols, Atmos. Chem. Phys., 15, 10723–10776, https://doi.org/10.5194/acp-15-10723-2015

ATTO – Amazon Tall Tower Observatory (2023). https://www.attoproject.org/pt/

Large Scale Biosphere-Atmosphere Experiment (LBA-ECO) (2023). https://daac.ornl.gov/cgi-bin/dataset_lister.pl?p=11

Sierra, J.P., Espinoza, J., Junquas, C., Wongchuig, P.J. S., Moron, V., Fita, L., Arias, P.A., Schrapffer, A., Pennel, R. (2023) Impacts of land-surface heterogeneities and amazonian deforestation on the wet season onset in southern. Climate Dynamics.

# Chapter 7
# Fast-Time Large-Scale Simulations for Unmanned Air Traffic Airspace Capacity Assessment

**Milan Rollo, Přemysl Volf, Lukáš Koranda, and Jan Straka**

## Nomenclature

| | |
|---|---|
| 2D | Two-dimensional |
| AED | Automated external defibrillator |
| ATM | Air traffic management |
| DAA | Detect and avoid |
| ICAO | International Civil Aviation Organization |
| KPI | Key performance indicator |
| UAS | Unmanned aerial system |
| UTM | Unmanned traffic management |

## 7.1 Introduction

Numerous research activities are currently being performed with the goal to design a U-space system that meets the current demand of all airspace users, even at peak times and locations, but also enables future growth. To be able to facilitate such a design, we must determine the key variables that describe the quality of such a system in order to be able to measure its performance and improve it by optimizing these indicators in future solutions.

We can draw inspiration from the existing key performance indicators (KPIs) defined for manned air traffic. Those are outlined in the ICAO's *Manual on Global Performance of the Air Navigation System* (Mobility 2020). Some of the indicators

M. Rollo (✉) · P. Volf · L. Koranda · J. Straka
AgentFly Technologies, Prague, Czech Republic
e-mail: rollo@agentfly.com; premysl.volf@agentfly.com; lukas.koranda@agentfly.com; jan.straka@agentfly.com

 
T. H. Karakoc et al. (eds.), *The Future of Electric Aviation and Artificial Intelligence*, Sustainable Aviation, https://doi.org/10.1007/978-3-031-62094-2_7

may be less relevant or important in the unmanned aviation domain, but capacity is one of the indicators that are. Others may be the safety of the operations, flight efficiency, or environmental and social impact. The relative importance of these areas and their mutual relationship may differ for unmanned air traffic compared to manned.

The airspace capacity is mainly affected by safety regulations that define minimal separation distances between aircraft. The air traffic management (ATM) traffic minimal horizontal separation is 5 NM. That is too prohibitive separation minima for urban traffic and smaller unmanned traffic management (UTM) aircraft. There are proposed different separation variants: static separation with smaller minima of 0.1 NM horizontally and 100 ft vertically (Bosson and Lauderdale 2018), dynamic separation based on aircraft class (performance and equipment type), or to implement detect and avoid (DAA) algorithms directly to aircraft.

The ground infrastructure also affects airspace capacity. The number and positions of vertiports, their throughput, or the limitations of their use due to outside factors like adverse weather greatly impact the number of aircraft operating in a given space.

There are established methods like Dynamic Density (Laudeman et al. 1998) or CAPAN (EUROCONTROL 2003) that are used to measure ATM airspace capacity based on ATCO's ability to control aircraft in the given airspace—communicating with aircraft and maintaining traffic safety. These methods cannot be directly reused for UTM because the unmanned traffic usually occupies the uncontrolled airspace that is not separated into individual sectors and there are no human controllers whose workload would be a limiting factor in the airspace capacity.

The UTM capacity can be also limited by the capacity of the communication spectrum. Urban areas do not always allow the use of a direct radio link to a drone, and drones have to rely on the existing communication networks. This can be a problem in highly populated and utilized areas. Based on a sampling-based estimation by Sedov and Polishchuk (Sedov et al. 2017), the communication capacity can be exceeded.

## 7.2 Capacity Assessment Using Fast-Time Simulation

There are multiple approaches related to airspace capacity described in Sect. 7.1. The airspace capacity is studied in this chapter as the maximum number of aircraft fitting into a given area and operating safely. The theoretical maximum is very close to filling the airspace exactly with "cylinders" of the aircraft safety zone size. In the case of a 2D plane (all aircraft flying at the same altitude), airspace is filled with circles. Such a theoretical maximum does not help evaluate the maximum possible traffic. Aircraft trajectories would be highly complex, hard to execute, and prolonged, and the number of conflicts would be very high. Thus, an important part of the capacity assessment is finding reasonable traffic restrictions to reach realistic flight trajectories and capacity.

An iterative approach from simple to complex traffic scenarios is used to understand the influence of parameters on metrics and capacity. The simulations are performed using AgentFly simulation platform (Rollo et al. 2017).

### 7.2.1 Assumptions

The fast-time simulation uses a model of reality. The model's details depend on the simulation's purpose, scale, and computational demandingness. Finding the capacity of the airspace requires running a large-scale simulation with hundreds to thousands of aircraft. Such a scale needs a simplified model to perform multiple simulations in a reasonable time frame.

The following assumptions are made to simplify the simulation model. Future studies can remove some of these assumptions to provide more realistic data. The airspace is assumed as single-layer airspace with a single-flight level. The free-flight trajectory is used to represent universal airspace utilization. The ideal trajectory execution is assumed for the simulation to prevent the increase in complexity by flight imprecisions that lead to multiple replannings. The capacity of takeoff and landing areas is assumed as unlimited because it is not intended to evaluate infrastructure in the scope of this work.

### 7.2.2 Straight-Line Traffic

The simplest scenario helps us understand the fundamental relationship between parameters and metrics. Generic traffic connects two waypoints by a direct straight trajectory (see Fig. 7.1). Aircraft planned start time is generated using equal time intervals. Increasing traffic leads to shortening the interval between planned starts.

When DAA is not allowed, aircraft follow one after the other in a straight trajectory. Maximum capacity is given by the safety zone's radius and the aircraft's speed. Figure 7.2 presents the number of planned to the actual number of aircraft. The chart shows all aircraft can be started at the moment when one aircraft follows the previous at the size of the safety zone. Once the limit is reached, no more aircraft can be started at the planned start time and needs to be delayed. The chart depicts multiple safety zones' sizes and their impact on the total capacity.

Figure 7.3 shows the ratio of the number of planned and the actual number of aircraft, and delays of aircraft. When the corridor capacity is not reached, no start delay is applied. Once the capacity is reached, the start delay gradually increases.

When DAA is allowed, but delaying start is not allowed, aircraft are forced to expand into surrounding airspace to reach the final waypoint. Initially, aircraft are flying in denser direct flow as in the previous scenario. Once the initial flow reaches its capacity, aircraft use DAA and form parallel flows on both sides of the initial

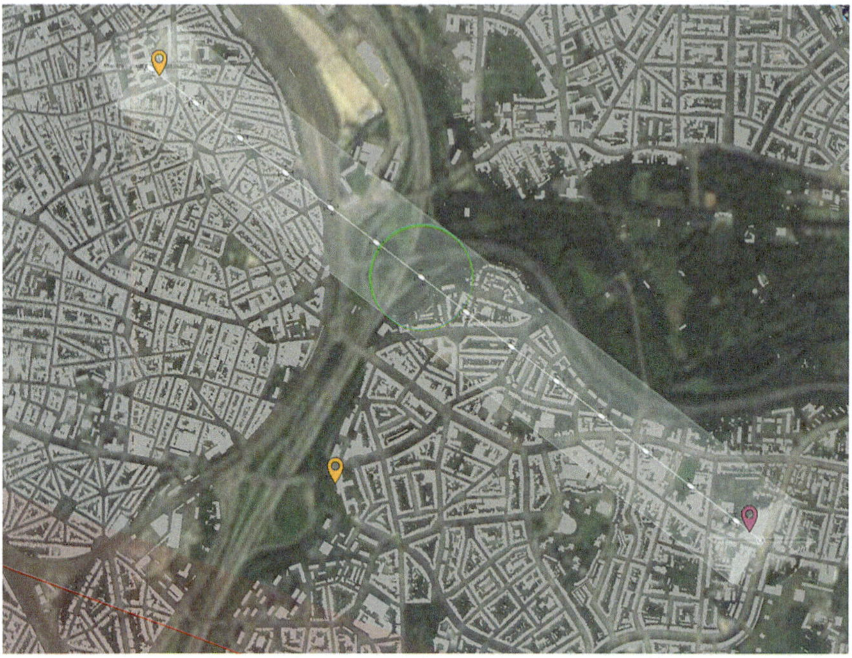

Fig. 7.1 Straight-line scenario connecting two waypoints

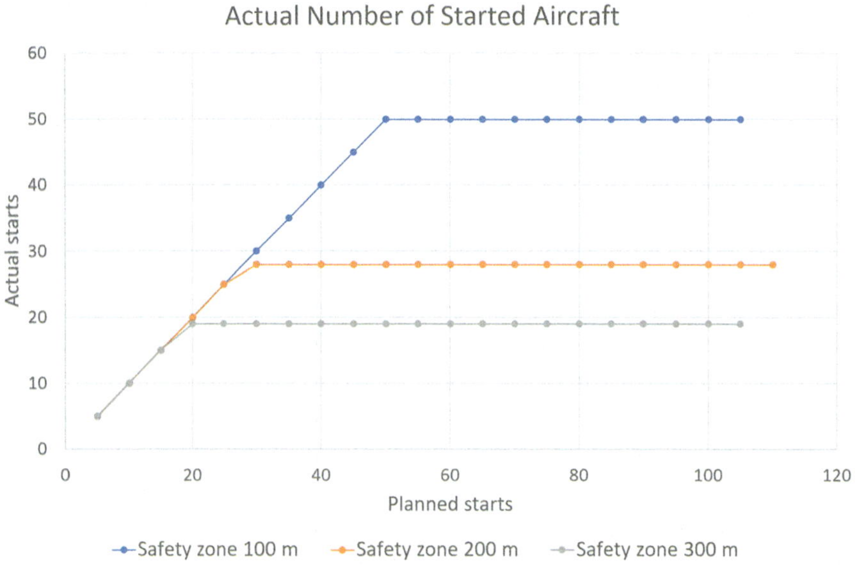


Fig. 7.2 Actual number of aircraft for different safety zones' sizes

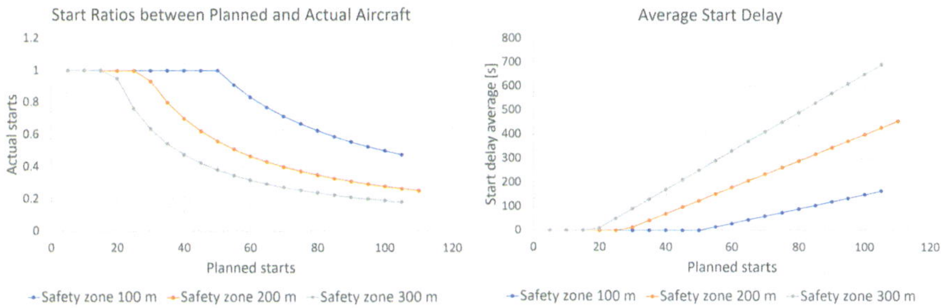


**Fig. 7.3** Ratio of the planned and actual number of aircraft for different safety zones' sizes on the left. Average start delay for different safety zones' sizes on the right

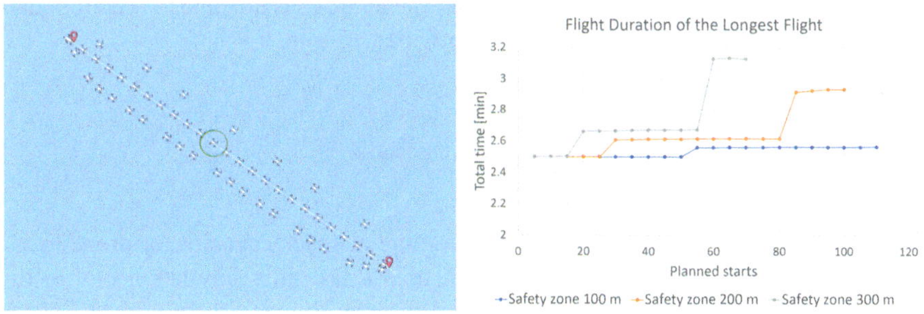


**Fig. 7.4** Side flows for increased traffic on the left. Extension of trajectory length for additional flows on the right

flow, and as traffic is further increased, more parallel flows are formed. The flight trajectory length is extended with additional flows (see Fig. 7.4).

### *7.2.3 Realistic Scenario: Prague*

Simulations of the city of Prague cover a large area of city and its surroundings. City center airspace is without restriction, except for Prague Castle's no-flight zone. The scenario traffic consists of multiple different types of flights. Medical deliveries between hospitals, stock supply flights from storage to hospitals and vertiports in city center, and medical flights to the near surroundings of hospitals (AED), and random flights around the city. Some of the flights are followed by a return flight to the start position.

Figure 7.5 (left) presents the number of actually started aircraft. Increasing traffic is possible due to sufficient airspace for performing DAA. Application of DAA and start delay allows us to increase traffic but also increase total flight duration. At some moment, the arrival time of priority flights may be too late. Figure 7.5 (right) presents a scenario using a maximum of 15 min for start delay. The capacity limit

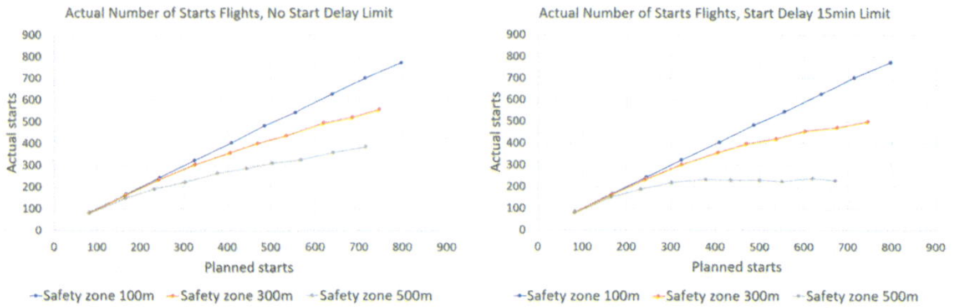


**Fig. 7.5** Left: actual number of started flights, no start delay limit. Right: the actual number of started flights, start delay 15 min limit

is reached for the 500 m safety zone because the airspace is not able to accommodate more flights without delaying them by more than 15 min.

## 7.3 Conclusions

The chapter presents the results of assessing the airspace capacity for drone traffic over urban areas. The presented approach uses a fast-time simulation tool to run many scenarios with increasing traffic to determine the maximum capacity of the airspace.

Initially, the methodology was proposed to define capacity and how to measure it. Elementary scenarios focusing on a single aspect of the capacity were created to fully understand the behavior of a large number of drones. Complex scenarios were simulated to study multiple combined aspects emulating future traffic in urban areas. Variation of parameters shows a major impact on capacity and the importance of simulations to assess the capacity for specific cities, environments, scenarios, and traffic.

Future work will focus on additional scenarios and parameter values. New methods of traffic management and detect and avoid algorithms will be used in simulations. Integration with other UTM components (e.g., C2C systems and drone operators) improves the fidelity of the simulation and the quality of capacity assessment.

**Acknowledgments** The research was carried out within the SAFIR-Med (Safe and Flexible Integration of Advanced U-space Services for medical Air Mobility) project funded by the European Union's Horizon 2020 research and innovation program under grant agreement no. 101017701 under the topic "SESAR-VLD2-03-2020: U-space capabilities and services to enable Urban Air Mobility."

## References

Mobility, F. (2020). Concept of Operations, v1. 0. Federal Aviation Administration (FAA): Washington, DC.

Bosson, C., and Lauderdale, T.A. (2018). Simulation Evaluations of an Autonomous Urban Air Mobility Network Management and Separation Service. In 2018 Aviation Technology, Integration, and Operations Conference AIAA AVIATION Forum. (American Institute of Aeronautics and Astronautics).

Laudeman, I.V., Shelden, S.G., Branstrom, R., and Brasil, C.L. (1998). Dynamic Density: An Air Traffic Management Metric.

EUROCONTROL EXPERIMENTAL CENTRE (2003). Pessimistic Sector Capacity Estimation.

Sedov, L., Polishchuk, V., and Bulusu, V. (2017). Sampling-based capacity estimation for unmanned traffic management. 2017 IEEE/AIAA 36th Digital Avionics Systems Conference (DASC). https://doi.org/10.1109/dasc.2017.8101995.

Rollo M., Selecký M., Volf P. (2017) Simulation of UAS integration into shared airspace for validation of impact on ATM systems, 2017 Integrated Communications, Navigation and Surveillance Conference (ICNS)

# Chapter 8
# On the Technological and Operational Aspects of the Greening of Aviation

L. M. B. C. Campos and J. M. G. Marques

## 8.1 Introduction

The emerging technologies (EFACA 2023; PARE 2019) setting path toward the greening of aviation (EC 2011; Zaporozhets et al. 2021) are considered for four aircraft classes: (1) battery power for small short-range aircraft: for example, electric vertical take-off and landing (e-VTOL) for urban air mobility (UAM); (2) fuel cells combined with turbines: for example, hybrid turboelectric power for regional aircraft; (3) hydrogen–fueled gas turbines: for example, jet-powered narrow-body airliners; and (4) sustainable aviation fuels (SAFs): for example, long-range widebody jetliners.

## 8.2 Battery Power for e-VTOL

The e-VTOL are complements/competitors/successors to helicopters with pros (+) and cons (−) for urban air mobility, with pros: (i) zero local emissions, although not elsewhere: mining of rare earth raw materials, production, and disposal of batteries, and production methods. (ii) Possibly lower noise and better community acceptance depending on the design. (iii) Possibly lower cost, depending on many factors. (iv) Many novel configurations possible due to low speed and short range. Cons are (v) weight and volume of batteries for the whole flight: no weight reduction by

L. M. B. C. Campos
Universidade de Lisboa, Instituto Superior Técnico, Lisbon, Portugal
e-mail: luis.campos@tecnico.ulisboa.pt

J. M. G. Marques (✉)
Universidade de Évora, Escola de Ciências e Tecnologia, Évora, Portugal
e-mail: jmgmarques@uevora.pt



T. H. Karakoc et al. (eds.), *The Future of Electric Aviation and Artificial Intelligence*, Sustainable Aviation, https://doi.org/10.1007/978-3-031-62094-2_8

fuel consumption. (vi) Lower speed, less range, and smaller payload to overall weight ratio than conventional helicopters. (vii) Less capability for autorotation emergency landing and flight in adverse weather. (viii) Need for rapid charging infrastructure at vertiports.

The operating conditions and limits include (i) safety of flight among tall buildings, including wind wake effects, may require better knowledge of "urban weather". (ii) Inability to overfly storms like aircraft requires the start of flight with safe weather prediction until landing. (iii) Limited operations in icing, fog, and rain may restrict yearly usage depending on location. (iv) Air traffic management (ATM) for low-altitude flights with urban obstacles may require new developments. (v) Limited autorotation capability for emergency landing may impose high safety requirements on overflying urban or populated areas. (v) Much fewer restrictions in rural, remote, or less populated areas with fewer obstacles: the opposite of UAM.

The vertiports and related infrastructure's possible locations include (i) existing heliports, with rapid electric charging infrastructure added. (ii) Roof-top heliports of private corporations may or may not be open to public use due to security concerns. (iii) New vertiports may face planning constraints: land area (will rolling landing may be required for emergencies?) and neighborhood noise. Also, a vertiport may cost a few million: Who pays? (i) Local authorities may not invest in "transport for the wealthy". (ii) Operators may not afford vertiports (airlines do not afford airports). (iii) Vertiports may need to be self-sustaining like airports: landing fees plus shop revenue. (iv) Will vertiports generate enough traffic or shopping revenues?

Certification rules worldwide: (i) On certification rules, the European Aviation Safety Agency (EASA) is the most advanced in the West and more stringent than the Federal Aviation Administration (FAA). (ii) At least EASA and FAA should harmonize rules to give designers uniform criteria. (iii) Rules applicable in China are little known on how they apply to flight operations already approved. (iv) Certification may take 4–5 years and cost 1 billion (not 2 years and 100 million as some claim). (v) Safety levels may be similar or the highest (EASA) in aviation because of low-altitude overflight of urban areas. (vi) Of the more than 100 e-VTOL projects by start-ups, about 4–6 may reach certification.

The overall cost components include (i) electricity per mile may be low, depending on local electricity rates. (ii) Cost of replacement of batteries (limited life not helped by fast charging for quick turnaround) may be comparable (or higher) than electricity costs. (iii) Plans for autonomous flying may take more than a decade (compared with cars in two dimensions)—pilot qualified for vertical flights may be required. (iv) Landing fees at vertiports. (v) Development costs need to be amortized over (long? or high rate?) production for sale cost of a few million. (vi) Maintenance may be less expensive than a helicopter, except for battery replacement. The cost is likely to be far higher than "Uber taxi" and somewhere between lightplane and helicopter, and closer to the latter.

The market prospects—"air taxi" is hardly feasible any time soon: (i) the number of car movements in a day in a large city may be larger than the number of aircraft flights worldwide in a year. (ii) ATM for millions of flights simultaneously over a city area? (iii) Not a solution to "surface traffic congestion," nor "Uber-like" transport fares. More likely early market is high-value local air transport: (i) Passenger transport between airports and city centers: airlines may want to expand existing helicopter services. (ii) Medical supplies: organs for transplant, urgent medicines and equipment, and emergency medical care following accidents: many hospitals already prepared with helicopter pads. (iii) Surveillance of remote facilities and agricultural tasks: less operational restrictions than flying over urban areas. (iv) Police and law enforcement: should foster public acceptance. (v) A gradual rise in utilization and importance should be expected.

## 8.3 Hybrid Turboelectric Propulsion

For example: regional airliner ATR 72 in class: 90 seats, 1000 km range; other subclasses: 10–12 seats, 400 km range, 30–50 seats, 600 km: both less demanding: (i) Turboprop sized for the most efficient cruise—longest stage of flight. (ii) Hydrogen fuel cell (HFC) for supplementary power for take-off and climb, and acceleration. (iii) Reverse possible: HFC for cruise and turbine for take-off, but not advisable, HFC is better for low emissions at low level. (iv) HFC at low altitudes emits no $CO_2$, $NO_x$, or particles, only wastewater. (v) Wastewater from HFC can be fed to the turbine engine to increase power. (vi) Using only HFC and battery power below 900 m altitude would ensure high local air quality (LAQ) at airports. (vii) Hybrid system must cope with failed landing and turn around, diversion to another airport, and could provide back-up for in-flight failure.

The performance required of HFC: (i) Battery power not suitable for regional aircraft with lithium-ion batteries, and perhaps not even with solid-state batteries: the latter best suited to improve the performance of UAMs. (ii) HFCs have a much better power-to-weight ratio than batteries but still fall short by a factor of 2–3 relative to aeronautical needs. (iii) Absolute power needed for regional aircraft at 1–2 MW per engine at the limit of HFC. (iv) Most of the weight of HFC is in the "plant": cooling and safety systems, hydrogen and oxygen supply, and other systems that need to be "miniaturized" by integrated design. (v) Safety and operating requirements are more stringent and critical in aeronautics than in surface transport (cannot stop at the roadside).

Also, current HFC experience with land vehicles does not cover operation at altitude: (i) for example, oxygen supply to HFC at 2 bar requires a compression ratio of 2:1 at sea level, 5:1 at 25,000 ft, and 7:1 at 35,000 ft, requiring a larger compressor. (ii) Cooling of the HFC by a liquid requires a heat exchange with atmospheric air, whose density decreases by a factor of 0.37 at 25,000 ft and 0.33 at 35,000 ft, requiring a larger radiator or heat exchanger. (iii) Operation at altitude does not help reduce plant mass and bulk and may influence safety issues. (iv) HFCs using current technology like permeable membrane (PEM) operate at "safer" low temperatures but have higher cooling requirements due to small temperature differences. (v) Solid oxid (SOF) HFC are at a lower TRL level and operate at higher temperatures with more efficient cooling, but pose new potential safety issues. (vi) Certification of HFC for aircraft is a major issue still partially open.

Market prospects for hybrid regional aircraft—retrofit of existing aircraft replacing fossil-fueled engines by HFC is possible with large emission reductions: (i) (A) 600 kW is enough for smaller 6–12 seats single, (B) 1 MW is needed in the 20–40 category, and (C) up to 2 MW for the 70–90-seat category. (ii) This assumes single-engine (A) or twin-engine (A, B, C) aircraft using parallel or series hybrid turboelectric propulsion with current technology PEM HFCs. (iii) Higher performance SOF HFC may be available only a generation later. (iv) The conversion of existing regional aircraft into HFC or hybrid propulsion is feasible in the near term and could make a significant reduction in emissions (40–60%). (v) This could be the precursor for regional aircraft designed from the outset for HFC or hybrid propulsion with higher emissions reductions (up to 80%).

## 8.4 Hydrogen-Fueled Narrowbody Jetliners

The main benefit of hydrogen versus fossil fuels is that it does not produce $CO_2$, although there are other emissions such as the combustion at higher temperatures can produce more NO*x*, requiring special attention. The water vapor is not a greenhouse gas (GHG) at low altitude but is at cruise altitudes in the form of contrails that can be countered in several ways: (i) Since less than 10% of the flights produce contrails, changing altitude eliminates them, with some ATM implications in areas of dense traffic. (ii) Treating hydrogen and optimizing combustion to produce fewer particles leaves fewer cores for the formation of ice crystals that constitute contrails. (iii) Trapping water vapor in the exhaust, cooling it into a liquid state, and injecting the water into the combustion chamber increases thrust while eliminating contrails, requiring more complex engine architecture (WET engine). (iv) The last solution is the best if technical and complexity implications can be mastered.

The hydrogen production and storage: (i) Hydrogen is abundant in water, but extraction by electrolysis requires more energy than is recovered by combustion. (ii) Hydrogen is a clean fuel if produced from a clean energy source: wind farms or solar panels. (iii) A large part of renewable energies is lost because they cannot be stored and are not used immediately: more sun is available in summer, but more heating is needed in winter. (iv) Using surplus clean energy to produce hydrogen is equivalent to energy storage for later use.

Hydrogen has higher (×4) energy content per unit weight than fossil fuels and much lower (×1/13) energy content per unit volume because it is the lightest element and gas: (i) Use of compressed hydrogen at 700 bar like in land vehicles not practical in aircraft: requires very heavy tanks with long filling times and slow turnaround. (ii) The alternative of liquid hydrogen (LH) requires very low temperatures (less than 20 K) and still occupies three times the volume of fossil fuel for the same energy.

The LH in aviation—on board the aircraft: (i) The whole LH supply system must be kept below 20 K. (ii) LH must be vaporized for combustion. There are safety issues on board: (i) Hydrogen is the lightest element, with the smallest atoms (one proton and one electron) and molecules (two atoms) and passes through most junctions: tubes must be welded. (ii) Hydrogen is corrosive, and steel tubes need protective coating. (iii) Hydrogen is explosive in a wide range of mixture ratios with air and must be vented rapidly, which is possible since it is the lightest of all gases. (iv) Hydrogen should be stored onboard preferably far from passengers: aft of the rear pressure bulkhead rather than in under fuselage hold vulnerable in belly landing. (v) Hydrogen has no smell or color and a leak can be detected by sensors at appropriate location (Where? Modeling?).

The safety applies all along the LH supply chain including refueling: (i) Refueling on the tarmac near the passenger terminal is perhaps not advisable. (ii) Preferable: replacement of "empty" by "full" LH tanks. (iii) Storage life of "full" LH tanks limited by leaks. (iv) If the storage time is approached, LH tanks must be returned to storage for "refilling."

## 8.5 SAFs for Widebody Jetliners

(i) The low energy-to-volume ratio of LH (one-third of fossil fuels) limits their use to narrow body jetliners up to 150 passengers and 1000–2000 nm range. (ii) The need for large LH tank volume may be difficult to accommodate in a conventional fuselage-wing configuration: not below the fuselage for safety, not enough volume in the wing, too long fuselage aft of rear pressure bulkhead. (iii) The large internal volume of a flying wing or blended wing body (BWB) configuration may be more suited to accommodate the large volume of LH tanks, subject to packaging issues (Cylindrical or spherical tanks? Supporting structure?). (iv) For a long-range widebody, say, 300 passengers and 16,000 km range, the volume of LH tanks would be prohibitive in terms of drag penalty, and a feasible fuel must have energy density per unit volume that approaches that of fossil fuels. (v) SAFs, either synthetic or biofuels, are not emission-free but are carbon-neutral if the $CO_2$ produced by combustion is matched by $CO_2$ taken out of the atmosphere in their production.

The 100% drop-in SAF: (i) Current certification rules allow blending with fossil fuel up to 50% SAF. (ii) Certification for 100% SAF will come with accepted standards for 100% SAF (as they already exist for jet fuel). (iii) Airbus and Boeing committed to making all their aircraft 100% SAF-compatible by 2030. SAF are ideally "drop-in" fuels that do not require any change in (i) aircraft fuel system, (ii) combustor design of jet engine, and (iii) airport fuel infrastructure. Fuel in aircraft is not only used for combustion but may have lubricant and other functions, needing consideration for SAFs of (i) long-term effects on seals, joints, and fuel distribution; (ii) flammability, volatility, and other safety aspects; (iii) full interchangeability with fossil fuels and among SAFs with different production paths; and (iv) compatibility with combustors and onboard and airport fuel system.

SAF availability in quantity—SAFs apply to all classes of aircraft but are most critical for long-range widebodies (LRW): (i) There is no alternative short-term solution to reduce emissions for LRW. (ii) "Long-range" (over 2000–3000 mm) flights consume 60–80% of all fossil fuels. (iii) Huge quantities of SAF are needed possibly in competition with other sectors. (iv) Main challenges: available feedstocks, energy, and facilities for production, final cost.

Feedstocks: (i) The most readily available feedstocks (cooking oil, urban waste, etc.) are not sufficient. (ii) Agricultural land suitable for food production should not be used for biofuels, only other land. (iii) There is a wide variety of feedstock, for example, algae blooms, some with high productivity per unit area, but very low efficiency of conversion to fuel, and immature processing technology. Energy consumption: (i) most promising SAFs are e-fuels combining hydrogen with $CO_2$ from the atmosphere; (ii) hydrogen production is energy intensive and requires 2–4 times the currently available electrical energy supplies: vast new wind and solar panel farms needed (or thermonuclear fusion reactors?).

SAF production process: (i) The Fisher–Tropf process developed in Germany during World War II and other alternatives require large investment in facilities. (ii) The investment is justified only with guaranteed long-term production prospects. (iii) Current SAF production is less than 1% of fossil fuel and is mostly sold out for several years, sometimes even before the facility is built and production starts. (iv) A very large ramp-up of production capacity is needed involving all stages: (a) feedstocks, (b) energy supply, and (c) refineries. SAF cost relative to fossil fuels: (i) SAFs currently cost 3–5 times more than fossil fuels. (ii) Fuel accounts for about 50% of airline costs. (iii) 100% SAF at these prices would double or treble airline costs and ticket prices. (iv) Moderate increase (20–30%) in ticket prices would require much lower SAF cost, together with vastly increased production. (v) $CO_2$ emission taxes (Valdés et al. 2021) can be seen as incentive or punitive but are not a technology solution and may limit the affordability of air travel. (vi) The alternative of carbon offsetting like planting trees exists to some extent.

## References

EC (2011) Flightpath 2050 Europe´s Vision for Aviation: Report of the High Level Group on Aviation Research, Publications Office of the European Union, EUR 098 EN, Luxemburg.

EFACA (2023) EC Project Reference: 101056866, accessed 06-2023, https://efaca.eu/

PARE (2019) Coordination and support Action 2017-1010, EC Project Reference: 769220, accessed 06-2023, https://www.pareproject.eu/publications

Valdés RMA, Comendador VFG, Campos LMB (2021) How Much Can Carbon Taxes Contribute to Aviation Decarbonization by 2050, *Sustainability*, *13*, 1086. doi:https://doi.org/10.3390/su13031086

Zaporozhets O, Isaienko, V, Synylo, K (2021) PARE preliminary analysis of ACARE FlightPath 2050 environmental impact goals, CEAS Aeronautical Journal 12:653–667. doi:https://doi.org/10.1007/s13272-021-00525-7

# Chapter 9
# Review and Analysis of Integration Properties of an Aircraft with a Hybrid Power Plant

**Vasyl Loginov, Oleksii Pushylin, and Oleksandr Tsaglov**

## Nomenclature

| | |
|---|---|
| ACARE | Advisory Council for Aviation Research and Innovation |
| EFACA | Environmentally Friendly Aviation for all Classes of Aircraft |
| EM | Electric motor |
| HTEPP | Hybrid turbo-electric power plant |
| GE | General Electric |
| IATA | International Air Transport Association |
| ICAO | International Civil Aviation Organization |
| MCU | Motor control unit |

## 9.1 Introduction

The perspective of increasing efficiency through evolutionary approaches is already reaching the limits of development. This means that radical changes in conceptual and technological thinking are needed when creating new models of aviation technology, taking into account the lack of environmental impact. One of these methods is the use of electric power plant in an airplane, helicopter, or UAV. Many aviation companies and organizations in the world have new development programs regarding electric onboard units (SAFRAN Press Kit 2019; IATA 2021). Most of the

V. Loginov (✉) · O. Tsaglov
Programs and Projects Department, FED JSC, Kharkiv, Ukraine
e-mail: aitsaglov@fed.com.ua

O. Pushylin
Ground Application Installations Department, Ivchenko-Progress SE, Zaporizhzhia, Ukraine
e-mail: pushilinae@ivchenko-progress.com

 
T. H. Karakoc et al. (eds.), *The Future of Electric Aviation and Artificial Intelligence*, Sustainable Aviation, https://doi.org/10.1007/978-3-031-62094-2_9

aircraft engine companies (e.g., SAFRAN, GE, P&W) have already started new programs to develop new components and units for engines (e.g., CFM RISE) powered by alternative fuels (biofuel, synthetic fuel, electric fuel), synthetic fuel mixed with kerosene, hydrogen gas, or hydrogen liquids. The results of hydrogen technology applications are expected by 2030 onboard aircraft.

To achieve the Flightpath 2050 goal (Flightpath 2050 2011), most aviation companies are combining for collaborative research and integral implementation of promising technologies. For example, a complex scientific and technical task was jointly solved by some organizations or companies during the implementation of the Horizon Europe programs (Horizon 2023; Clean Sky 2 2020) together. An analysis of modern research on the development of aircraft has shown that the use of hybrid technology when creating a power plant as part of an aircraft is the main method for achieving better performance (Kiruba 2021).

## 9.2 Problem Analysis and Research Methods

It is known that two main subsystems are distinguished in a modern aircraft: the airframe and the hybrid power plant. These subsystems have the greatest influence on the flight's technical and economic characteristics of the aircraft.

The hybrid power plant is a separate complex technical subsystem in terms of the complexity of physical processes occurring in flight, which determines the main flight and technical characteristics of the "flying vehicle" system. In order to reduce the impact of aviation on the environment, new methods and ways of creating hybrid power plants are based on the

- Application of battery and turboelectric schemes (hybrid technologies).
- Direct combustion of hydrogen or hydrogen mixes in the combustion chamber.
- Application of hydrogen fuel cells powering electric motors.

In this case, the choice of the main indicator or criterion for the integration of a hybrid power plant as part of an aircraft is a complex task. The analysis of methods and approaches shows that the integration of a hybrid power plant as part of an aircraft is a multicriteria and multidisciplinary task (Shmyrov 2020). Therefore, the problems of system analysis of the hybrid power plant as part of an aircraft are closely related to each other. The authors developed special mathematical methods, algorithms, and models, providing the use of a wide database, the ability to analyze the influence of a large number of factors, as well as optimization of parameters, taking into account alternative criteria and many limitations.

As is known, the creation of the airframe is currently carried out separately from the creation of a hybrid power plant. However, the creation of a single airframe with high technical characteristics is not a guarantee of high flight and technical characteristics with the intended hybrid power plant since the flight data of the aircraft always differ from the calculated ones. This is due to the appearance of new integrative properties of mutually influencing two subsystems: airframe and hybrid

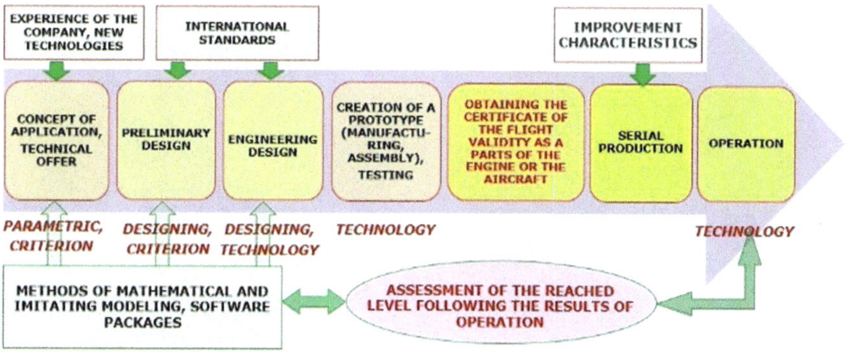


**Fig. 9.1** The stages of integration of the airframe and hybrid power plant in the life cycle of an aircraft

power plant. The combination of subsystems characteristics in their interconnection at the level of work processes determines the properties of the aircraft as a whole.

In order to obtain the characteristics of a hybrid power plant as part of an aircraft, the stages of integration are investigated: parametric, criterion, designing, and technology. The complexity and importance of technical decisions regarding the definition of integration properties can be seen in the example of the life cycle of an aircraft (Fig. 9.1). We see interconnected processes from the conception phase of an aircraft to the end of its operation.

Analysis results of modern technical solutions regarding the creation of a hybrid power plant (Schmuck 2019; Palladino 2021) show that the characteristics of electric motor (EM) are an important part of electric aircraft. Therefore, research and obtaining effective technical and economic characteristics of EM are an important part of the creation of hybrid turbo-electric power plant (HTEPP) (Moebs 2022; Marciello 2023).

## 9.3 Study of EM Design Characteristics

To evaluate the level of technical implementation of the future design, the characteristics of EM from different manufacturers were analyzed. Preliminary research was carried out to develop an advanced electric motor for a hybrid power plant. Issues on the key characteristics of electric motor, its control system, and software/hardware were considered. As a result of research, the possible configurations of an advanced EM and its characteristics are received (Fig. 9.2).

The features of the proposed EM are outrunner architecture, NbFeB permanent magnets, and specific rotor configuration with inner magnets location, which are obtained during multidisciplinary development (electromagnetic, thermal, and strength analysis in Ansys Electronics, Ansys Workbench, and custom programs).

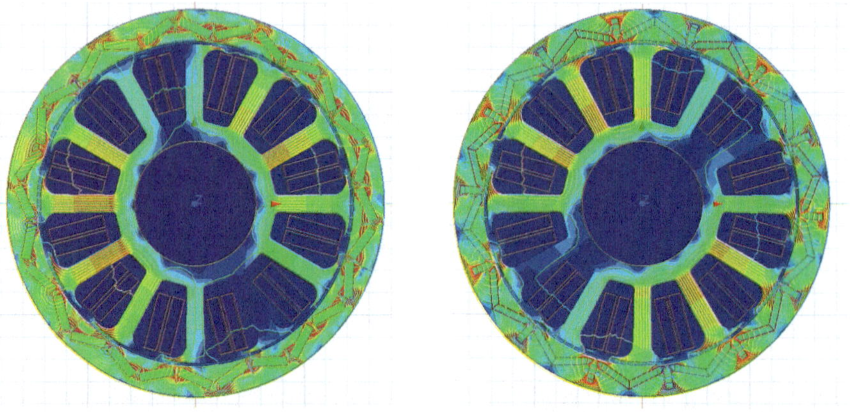

**Fig. 9.2** Study of specific rotor configuration of EM (magnetic inductance and field line distribution in electric motor section)

EM control systems (separate and integrated configurations) have state-of-the-art IGBTs and SiC MOSSFETs, redundant fault-tolerant design with a high specific power ratio, and multidisciplinary development (Altium Designer, Simulink, Ansys Electronics, etc.). Hardware and software are based on model-based development for airborne application (ARP4754, DO178, DO254, DO331 complies).

This analysis was carried out as a result of the joint work of JSC FED and Ivchenko-Progress SE in the EFACA project. Environmentally Friendly Aviation for all Classes of Aircraft (EFACA) is a project funded by the European Commission through the Horizon Europe Programme (grant agreement no. 101056866), inserted on the Climate, Energy and Mobility cluster. The EFACA intends to promote an aviation sector greener and environmentally friendly with the development of new technologies by using electric and hybrid thermoelectric propulsions and new sustainable fuels aiming at fighting climate change by 2050.

## 9.4 Conclusion

Based on the study of two prototypes of electric motors with different initial configuration, their results are obtained. When evaluating a prototype with output power 120 kW, operation speed 20,000 rpm, supply voltage 540 VDC, specific power rate $>$ 8 kW/kg, with rotor position sensor, the estimated weight is about 15 kg (EM + microcontroller). When evaluating a prototype with output power 150 kW, operation speed 20,000 rpm, supply voltage 800 VDC, specific power rate $>$ 12 kW/kg, with integrated sensorless control system, the estimated weight is about 12 kg (EM + MCU).

The volumetric characteristics of EM and the place of installation as part of the hybrid power plant will be specified in the following studies.

## References


SAFRAN (2019), "Safran and Aviation's Electric Future," Press Kit, 2019 Paris Air Show. – 36 p.

Clean Sky 2 (2020) Hydrogen-powered aviation, A fact-based study of hydrogen technology, economics, and climate impact by 2050. https://doi.org/10.2843/766989.

Electrified aircraft propulsion: powering the future of air transportation/edited by Kiruba Haran, Nateri Madavan, Tim C. O'Connell.: Cambridge, United Kingdom; New York, NY, USA: Cambridge University Press, 2021. – 317 p. https://doi.org/10.1017/9781108297684.

IATA (2021), "Aircraft Technology Roadmap to 2050," 2019. [Online]. Available: https://www.iata.org/en/programs/environment/technology-roadmap/. – Accessed August 2021.

Flightpath 2050 Europe's Vision for Aviation (2011). Report of the High Level Group on Aviation Research. – Luxembourg: Publications Office of the European Union, 2011. – https://doi.org/10.2777/50266.

Horizon (2023) Europe work programmes. – https://research-and-innovation.ec.europa.eu/funding/funding-opportunities/funding-programmes-and-open-calls/horizon-europe/horizon-europe-work-programmes_en. – Accessed 20.07.2023.

V. Shmyrov (2020), V. Loginov, S. Fil, A. Khaustov, O. Bondarchuk, A. Kalashnikov, G. Khmelnitskiy. The modernization concept of aircraft An-26 and An-140 based on the use of a hybrid power system. *Eastern-European Journal of Enterprise Technologies*. No.5. pp. 6–17. https://doi.org/10.15587/1729-4061.2020.212150

M. Schmuck (2019), J. Garche, St. Koller. Emerging Aviation Technologies – Progress in the Electrification of Aircraft/8th European conference for aeronautics and space sciences (EUCASS). https://doi.org/10.13009/EUCASS2019-851.

Palladino V. (2021), Jordan A., Bartoli N., Schmollgruber P., Pommier Budinger V., et al. Preliminary studies of a regional aircraft with hydrogen-based hybrid propulsion. AIAA AVIATION 2021 FORUM, Aug 2021, Virtual event, United States. – pp. AIAA 2021–2411. https://doi.org/10.2514/6.2021-2411.hal-03690045.

Moebs, N.; Eisenhut, D.; Windels, E.; van der Pols, J.; Strohmayer, A. Adaptive Initial Sizing Method and Safety Assessment for Hybrid-Electric Regional Aircraft. Aerospace 2022, 9, 150. https://doi.org/10.3390/aerospace9030150.

Marciello, V.; Di Stasio, M.; Ruocco, M.; Trifari, V.; Nicolosi, F.; Meindl, M.; Lemoine, B.; Caliandro, P. Design Exploration for Sustainable Regional Hybrid-Electric Aircraft: A Study Based on Technology Forecasts. – Aerospace 2023, 10, 165. https://doi.org/10.3390/aerospace10020165.

# Chapter 10
# Optimization of Quadcopter Energy Consumption: Insights from Wind Condition Analysis and Trajectory Planning

**Mohammed Edawdi and Marcin Żugaj**

## Nomenclature

UAV Unmanned aerial vehicle

## 10.1 Introduction

The quest to optimize energy consumption in quadcopters has been an ongoing endeavor in the field of aerospace engineering. In recent years, the increasing application of quadcopters for various tasks, from surveillance to package delivery, has necessitated the development of energy-efficient flight strategies. Such strategies could significantly extend the operational duration of quadcopters and reduce operational costs.

Studies such as (Argentim et al. 2013), (Nemati and Kumar 2014), (Fruncillo et al. 2022), (Abdulkareem et al. 2022), (Koch et al. 2019), (Liu et al. 2017), and (Abeywickrama et al. 2018) have made significant strides in creating dynamic models for quadcopters. These works, which range from simplistic kinematic models to more intricate dynamic simulations, have provided a solid foundation for understanding and predicting quadcopter behavior. However, the accuracy of these models in terms of energy consumption prediction often varies, with many studies identifying the need for more precise energy consumption models as emphasized by (Zhang et al. 2021) and (Prasetia et al. 2019).

M. Edawdi (✉) · M. Żugaj
Faculty of Power and Aeronautical Engineering, Warsaw University of Technology, Warsaw, Poland
e-mail: mohammed.edawdi.dokt@pw.edu.pl; marcin.zugaj@pw.edu.pl



T. H. Karakoc et al. (eds.), *The Future of Electric Aviation and Artificial Intelligence*, Sustainable Aviation, https://doi.org/10.1007/978-3-031-62094-2_10

Energy consumption in quadcopters is largely dependent on their operational strategies. (Hepp et al. 2018) and (Belge et al. 2022) presented valuable findings on the role of flight path optimization in energy conservation. Yet, these works mainly focused on reflective of real-world scenarios.

Furthermore, the impact of environmental factors, particularly wind conditions, on the energy consumption of quadcopters has been explored by several researchers (Thu et al. 2021), (Yacef et al. 2020). Their findings highlight how wind speed and direction can significantly influence the energy consumption of these unmanned aerial vehicles (UAVs). However, there remains a gap in the integration of these wind conditions into the trajectory planning for optimal energy use.

Our research was motivated by the desire to devise optimal flight paths and strategies for a lightweight quadcopter with the aim of minimizing energy usage. The groundwork for this project was laid by (Jacewicz et al. 2022), which offers an in-depth examination of the energy efficiency of a quadrotor dynamic model. Their proposed model, which is instrumental in our simulations, captures the impact of the drone's maneuvers on its energy consumption. It uses rotor thrust, torque, and drag force correction factors calculated using FLIGHTLAB software to accurately model quadrotor dynamics in forward flight.

In this chapter, we focus on further enhancing the value of this dynamic model. We explore its utility in studying the effects of varying wind conditions and different trajectories on the quadcopter's energy consumption. Our goal is to provide empirical evidence that will guide the planning and design of energy-efficient flight paths for quadcopters operating in diverse environments and weather conditions.

## 10.2 Methodology

A speed of 10 m/s (roughly 67% of the quadcopter's maximum speed) offers an optimal balance for our study. While it is true that higher speeds result in higher energy consumption, 10 m/s provides a substantial velocity for quick deployment without excessively increasing energy demand. This speed thus enables a more efficient energy usage while maintaining a reasonable operational performance, making it a suitable choice for conducting the remainder of our analysis.

### *10.2.1 Choice of Trajectory*

Trajectory 1 (Fig. 10.1) is a zigzag flight path in a square area of 1,000,000 $m^2$ (1000 m × 1000 m).

The quadcopter travels a distance of $\lambda$ (50 m) along the x-axis, then 1000 m along the y-axis, repeating this pattern 20 times.

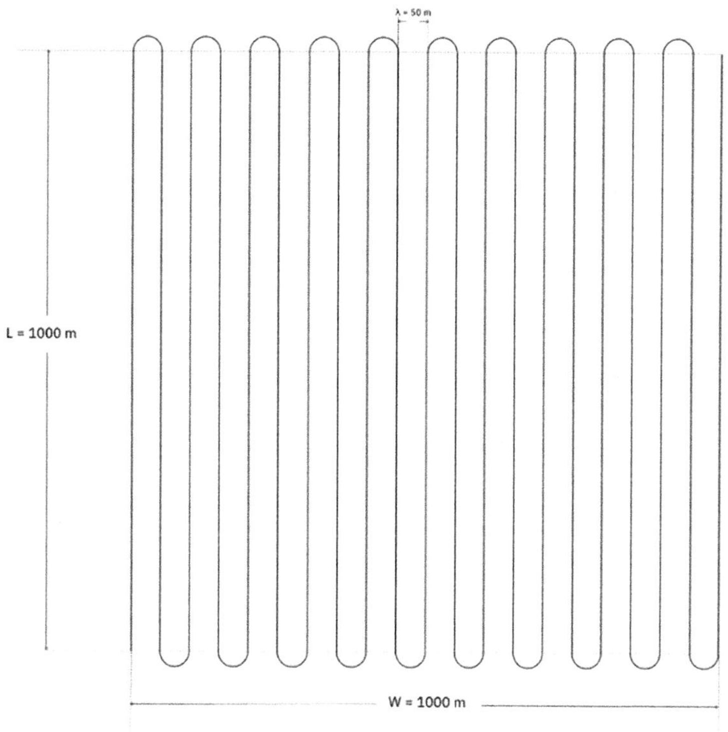


**Fig. 10.1** Trajectory 1

Trajectory 2 (Fig. 10.2) mirrors trajectory 1 but spans a rectangular area of 1,000,000 $m^2$ (500 m × 2000 m). The quadcopter alternates between positive and negative y-axis movements in a zigzag pattern.

Trajectory 3 (Fig. 10.3) also follows a zigzag flight pattern but within a rectangular area of 1,000,000 $m^2$ (2000 m × 500 m). The quadcopter alternates between positive and negative y-axis movements, completing 40 zigzags. All three trajectories assume the quadcopter moves in perfectly straight lines and makes instantaneous turns.

In determining the optimal trajectory for our energy consumption analysis under various wind conditions, we considered factors such as wind direction and intensity, path length, maneuverability, and wind variability. For zero-wind conditions, trajectory selection was flexible and based on considerations such as flight time or maneuver complexity. With headwinds, we favored trajectory 3 due to its longer path aligning with the wind direction. For crosswind and oblique wind conditions, trajectories 1 and 2 were deemed beneficial due to their frequent direction changes, necessitating more energy-demanding corrections.

Of particular interest, trajectory 1 proved to be an effective choice across all wind conditions. Its balanced mix of longitudinal and lateral movements allowed for a diverse range of wind exposures. The trajectory's 20 zigzags provided ample data on the quadcopter's energy management during direction changes against varying wind conditions.

**Fig. 10.2** Trajectory 2

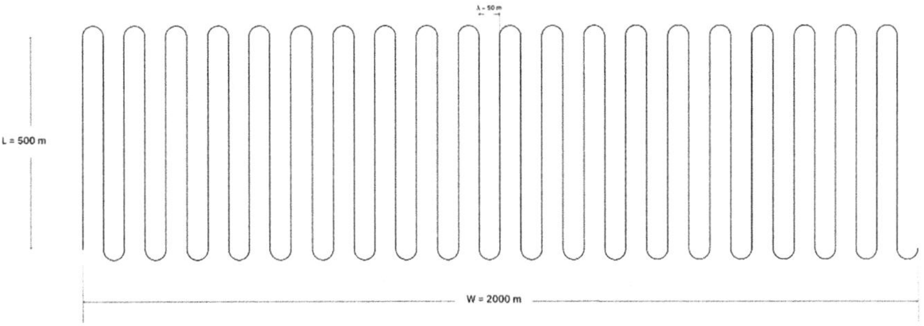


**Fig. 10.3** Trajectory 3

We emphasize, however, that the ideal trajectory selection will be contingent on the specific goals of a given study and the unique dynamics of the quadcopter in use. Nevertheless, our findings suggest that a balanced and maneuver-diverse trajectory, such as trajectory 1, can provide a comprehensive dataset for assessing the impact of various wind conditions on quadcopter energy consumption.

## 10.3 Results and Discussion

The data (Figs. 10.4, 10.5 and Table 10.1) clearly exhibits that both wind speed and wind direction play crucial roles in the energy consumption of quadcopters. An increase in wind speed from 0 m/s to 7 m/s corresponds with an increase in both engine and total energy consumption. The underlying reason for this observation can be logically attributed to the additional energy required to counter the increased wind forces.

Similarly, a change in wind direction from being parallel to the flight path (0°) to perpendicular (90°) also leads to an increase in energy consumption. This trend suggests that, when the wind direction is parallel to the flight path, it is possible for the quadcopter to utilize the wind to some extent, thereby reducing energy consumption. Conversely, wind perpendicular to the flight path likely necessitates additional energy to maintain course, leading to elevated energy consumption.

The discrepancy between engine energy consumed and total energy consumed is due to additional energy use attributable to onboard systems. This differential remains relatively stable across all wind settings, implying that these systems consume a consistent amount of energy irrespective of wind conditions.

These observations provide valuable insights for optimizing energy consumption in quadcopters. By factoring in prevailing wind conditions when planning a drone's flight path, it is feasible to minimize energy consumption and consequently extend the quadcopter's flight duration. This approach can enhance operational efficiency and potentially prolong the quadcopter's service life by reducing stress on the engine.

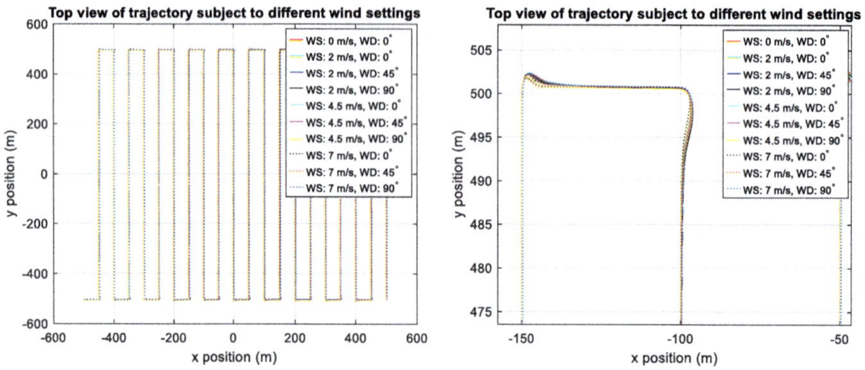


**Fig. 10.4** Top view of the trajectory followed

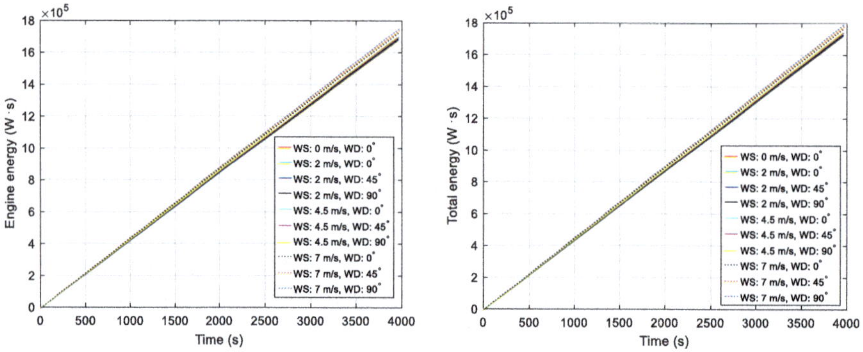


**Fig. 10.5** Total (left) and engine (right) energy consumption

**Table 10.1** Simulation results

| Wind setting | Engine energy consumed (W·s) | Total energy consumed (W·s) |
|---|---|---|
| WS: 0 m/s & WD: 0° | 1.6806*e* + 06 | 1.7241*e* + 06 |
| WS: 2 m/s & WD: 0° | 1.6834*e* + 06 | 1.7269*e* + 06 |
| WS: 2 m/s & WD: 45° | 1.6849*e* + 06 | 1.7284*e* + 06 |
| WS: 2 m/s & WD: 90° | 1.6860*e* + 06 | 1.7295*e* + 06 |
| WS: 4.5 m/s & WD: 0° | 1.6955*e* + 06 | 1.7391*e* + 06 |
| WS: 4.5 m/s & WD: 45° | 1.7020*e* + 06 | 1.7455*e* + 06 |
| WS: 4.5 m/s & WD: 90° | 1.7077*e* + 06 | 1.7513*e* + 06 |
| WS: 7 m/s & WD: 0° | 1.7251*e* + 06 | 1.7688*e* + 06 |
| WS: 7 m/s & WD: 45° | 1.7411*e* + 06 | 1.7848*e* + 06 |
| WS: 7 m/s & WD: 90° | 1.7550*e* + 06 | 1.7987*e* + 06 |

## 10.4 Conclusion

Our study addressed the critical issue of energy consumption in quadcopters, revealing important insights into how wind conditions impact energy usage. By conducting rigorous simulations using a dynamic model for reliable energy consumption estimation, we demonstrated that both wind speed and wind direction significantly affect the energy consumption of these unmanned systems.

Our results indicate that an increase in wind speed, as well as a shift in wind direction from being parallel to the flight path to perpendicular, results in higher energy consumption. This information can be crucial in planning efficient flight paths, especially for surveillance applications, to minimize energy usage and extend operational capabilities.

The consistent difference between engine and total energy consumed across all wind settings suggests that additional energy used for onboard systems remains relatively constant, irrespective of wind conditions. This understanding can help in developing more efficient systems and subsystems within the quadcopter to further optimize energy consumption.

## References


Abdulkareem, A., Oguntosin, V., Popoola, O. M. & Idowu, A. A., 2022. Modeling and Nonlinear Control of a Quadcopter for Stabilization and Trajectory Tracking. *Journal of Engineering (United Kingdom),* Volume 2022.

Abeywickrama, H. V., Jayawickrama, B. A., He, Y. & Dutkiewicz, E., 2018. Comprehensive Energy Consumption Model for Unmanned Aerial Vehicles, Based on Empirical Studies of Battery Performance. *IEEE Access,* Volume 6, pp. 58383–58394.

Argentim, L. M., Rezende, W. C., Santos, P. E. & Aguiar, R. A., 2013. PID, LQR and LQR-PID on a quadcopter platform. *2013 International Conference on Informatics, Electronics and Vision, ICIEV 2013.*

Belge, E., Altan, A. & Hacıoğlu, R., 2022. Metaheuristic Optimization-Based Path Planning and Tracking of Quadcopter for Payload Hold-Release Mission. *Electronics 2022, Vol. 11, Page 1208,* 4, 11(8), p. 1208.

Fruncillo, F., Federico, L., Cicala, M. & Citarella, R., 2022. Development and Validation of an Aeropropulsive and Aeroacoustic Simulation Model of a Quadcopter Drone. *Drones 2022, Vol. 6, Page 143,* 6, 6(6), p. 143.

Hepp, B., Zurich, E., Niessner, M. & Hilliges, O., 2018. Plan3D: Viewpoint and Trajectory Optimization for Aerial Multi-View Stereo Reconstruction.

Jacewicz, M., Żugaj, M., Głębocki, R. and Bibik, P., 2022. Quadrotor Model for Energy Consumption Analysis. *Energies,* 15(19), p. 7136.

Koch, W., Mancuso, R., West, R. & Bestavros, A., 2019. Reinforcement Learning for UAV Attitude Control.

Liu, Z., Sengupta, R. & Kurzhanskiy, A., 2017. A power consumption model for multi-rotor small unmanned aircraft systems. *2017 International Conference on Unmanned Aircraft Systems, ICUAS 2017,* 7. pp. 310–315.

Nemati, A. & Kumar, M., 2014. Modeling and control of a single axis tilting quadcopter. *Proceedings of the American Control Conference,* pp. 3077–3082.

Prasetia, A. S., Wai, R. J., Wen, Y. L. & Wang, Y. K., 2019. Mission-based energy consumption prediction of multirotor uav. *IEEE Access,* Volume 7, pp. 33055–33063.

Thu, A., Lupin, S., Oo, T. M. & Khaing, M. T., 2021. Comparing a Quadrotor Energy Consumption for Different Flight Trajectories in Windy Conditions. *Proceedings of the 2021 IEEE Conference of Russian Young Researchers in Electrical and Electronic Engineering, ElConRus 2021,* 1. pp. 2064–2066.

Yacef, F., Rizoug, N., Degaa, L. & Hamerlain, M., 2020. Energy-Efficiency Path Planning for Quadrotor UAV under Wind Conditions. *7th International Conference on Control, Decision and Information Technologies, CoDIT 2020,* 6. pp. 1133–1138.

Zhang, J., Campbell, J. F., Sweeney, D. C. & Hupman, A. C., 2021. Energy consumption models for delivery drones: A comparison and assessment. *Transportation Research Part D: Transport and Environment,* 1, Volume 90, p. 102668.

# Chapter 11
# Sustainable Aviation Fuels and Their Use for Long-Range Aircraft

**Renata Adami, Patrizia Lamberti, Maria Sarno, and Vincenzo Tucci**

## Nomenclature

| | |
|---|---|
| ASTM | American Society for Testing and Materials |
| IATA | International Air Transport Association |
| CAAFI | Commercial Aviation Alternative Fuels Initiative |
| RED | Renewable Energy Directive |
| GHG | Greenhouse gases |
| SAF | Sustainable aviation fuels |

## 11.1 Introduction

The European Green Deal (European Commission 2019) aims to drive the EU toward a sustainable and fair society. The goal is to reach in 2050 no net emissions of greenhouse gases (GHG) and improve the resource usage for an efficient and competitive economic growth. The decarbonization of the transport sector is a

R. Adami (✉) · M. Sarno
Department of Physics "E.R. Caianiello", University of Salerno, Fisciano, SA, Italy

Interdepartmental Centre NANO_MATES (Research Centre for NANOMAterials and nanoTEchnology), Fisciano, SA, Italy
e-mail: radami@unisa.it; msarno@unisa.it

P. Lamberti · V. Tucci
Interdepartmental Centre NANO_MATES (Research Centre for NANOMAterials and nanoTEchnology), Fisciano, SA, Italy

Department of Information, Electrical Engineering and Applied Mathematics, University of Salerno, Fisciano, SA, Italy
e-mail: plamberti@unisa.it; vtucci@unisa.it



T. H. Karakoc et al. (eds.), *The Future of Electric Aviation and Artificial Intelligence*, Sustainable Aviation, https://doi.org/10.1007/978-3-031-62094-2_11

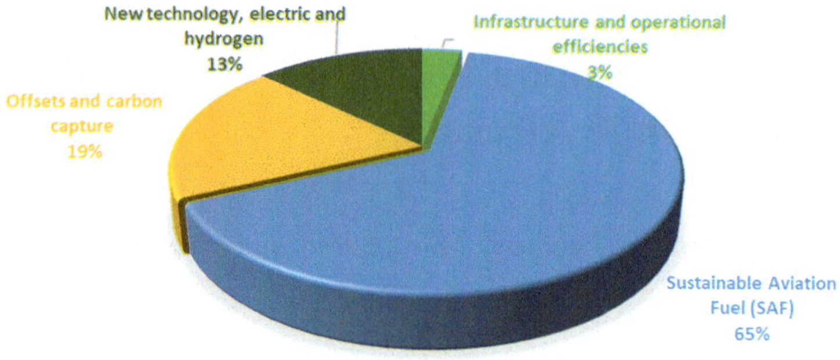


**Fig. 11.1** Contribution to achieving net-zero carbon in 2050

fundamental part of the plan. Within this sector, aviation is the most difficult to decarbonize. A target of reducing by 50% $CO_2$ emissions compared to the 2005 level target has been fixed by the International Air Transport Association (IATA).

The aviation industry's net-zero carbon emissions target is focused on delivering maximum reduction at source through the use of Sustainable Aviation Fuels (SAF), innovative new propulsion technologies, and other efficiency improvements, such as those to air traffic navigation (IATA 2022). New propulsion systems include battery- and turbo-electric technologies, as well as hydrogen combustion in turbines and fuel cells that power electric motors. Batteries can be applied in combination with conventional propulsion (turbo-electric aircraft) or hydrogen fuel cells (Misra 2017; Seitz et al. 2012; Ashcraft et al. 2011; Comincini 2018).

The usage of SAF could contribute around 65% of the reduction in emissions needed by aviation to reach net zero in 2050 (Fig. 11.1). This will require a massive production increase in order to meet the demand. The largest acceleration is expected to happen in the 2030s; as policy support becomes global, SAF becomes competitive with kerosene.

For small, lightweight aircraft, including Urban Air Mobility battery, power is suitable; for airlines up to 150 passengers and 2000-km-range, turbine power with liquid hydrogen may be a possible solution. For the intermediate case of regional aircraft, hybrid turbo-electric propulsion using turboprops and fuel cells is the solution.

For very large, long-range aircraft, hydrogen fuel is not viable due to large tank volume and high drag. The use of gas turbines to drive electric generators should provide several megawatt of electric power needed for distributed electric propulsion. Therefore, for very large aircraft and very long distances, low-carbon alternative fuels are at present the only viable replacement in the future for fossil fuels. They have reduced particulates, nitrogen, sulfur, and critical greenhouse gases and have the advantage of being "drop-in fuels" that do not require changes in aircraft, neither for the engines nor for the fuel tanks, and fuel infrastructure or airport facilities. The SAF considered are

- **Biofuels** from biomass or waste (cooking oils, fats) and **advanced biofuels** that are synthesized from solid feedstock, biomass like crops, or algae.
- **e-fuels,** which are synthetic fuels or synfuels, or synthesized from hydrogen and $CO_2$ taken from industrial, biomass, or direct-air capture.

## 11.2 Biofuels and Advanced Biofuels

To bring new fuels in commercial flights, a long and expensive route encompassing certification is needed, according to the normative fixed by the American Society for Testing and Materials (ASTM):

- ASTM D1655 Standard Specification for Aviation Turbine Fuels.
- ASTM D7566 Standard Specification for Aviation Turbine Fuel Containing Synthesized Hydrocarbons.
- ASTM D4054 Standard Practice for Qualification and Approval of New Aviation Turbine Fuels and Fuel Additives.

The entire path necessary to certify a new aviation fuel, that requires significant efforts and resources, is summarized in the ASTM D4054 guideline, whereas the legislative basis for use of advanced sustainable biofuels in Europe is the revision of the RED II (Chiaramonti and Goumas 2019; The European Parliament and the Council of the European Union 2018). The Commercial Aviation Alternative Fuels Initiative (CAAFI) reports that there are seven major fuel routes approved by the ASTM D7566 standard (Table 11.1):

- Annex A1 – Fischer-Tropsch fuels using solid biomass resources (FT-SPK)
- Annex A2 – Hydrogenated esters and fatty acids (HEFA) fuels derived from used cooking oil, animal fats, algae, and vegetable oils (HEFA-SPK)
- Annex – A3 Synthetic iso-paraffin from fermented hydroprocessed sugar (SIP-SPK)
- Annex – A4 FT fuels with aromatics using solid biomass resources (FT-SKA)
- Annex – A5 Alcohol-to-jet fuels produced from isobutanol (ATJ-SPK)
- Annex – A6 Catalytic hydrothermolysis synthesized kerosene (CH-SK, CHJ)
- Annex –A7 Hydroprocessed hydrocarbons (HH-SPK, HC-HEFA)

Moreover, in Annex A1 of ASTM D1655, the co-processing of renewable content with crude oil-derived middle distillates in petroleum refineries has been approved. This includes lipids (plant oils and animal fats) and Fischer–Tropsch biocrude (unrefined hydrocarbon content coming from the reactor), as co-processing of up to 5% by volume of these components as feedstocks in petroleum refinery processes.

CAAFI also reports that 6 routes are currently under approval process, while 15 are waiting to enter the process and a significant number of additional prospective pathways are currently being pursued by multiple entities but have yet to enter the

**Table 11.1** Processes pathway approved by ASTM

| Pathways processes | Feedstock | Date of approval | Blending ratio by volume | Commercialization proposals |
|---|---|---|---|---|
| Fischer–Tropsch synthetic paraffinic kerosene (FT-SPK) | Biomass (forestry residues, grasses, municipal solid waste) | 2009 | 50% | Fulcrum Bioenergy, Red Rock Biofuels, SG Preston, Kaidi, Sasol, Shell, Syntroleum |
| Hydroprocessed esters and fatty acids (HEFA-SPK) | Oil-bearing biomass, e.g., algae, jatropha, camelina, carinata | 2011 | 50% | World Energy, Honeywell UOP, Neste Oil, Dynamic Fuels, EERC |
| Hydroprocessed fermented sugars to synthetic isoparaffins (SIP-HFS) | Microbial conversion of sugars to hydrocarbon | 2014 | 10% | Amyris, Total |
| FT-SPK with aromatics (FT-SPK/A) | Renewable biomass such as municipal solid waste, agricultural wastes and forestry residues, wood and energy crops | 2015 | 50% | Sasol |
| Alcohol-to-jet synthetic paraffinic kerosene (ATJ-SPK) | Agricultural wastes products (stover, grasses, forestry slash, crop straws) | 2016 | 50% | Gevo, Cobalt, Honeywell UOP, Lanzatech, Swedish Biofuels, Byogy |
| Hydroprocessed esters and fatty acids plus (HEFA +) | Oil-bearing biomass, e.g., algae, jatropha, camelina, carinata | 2021 | 50% | Boeing |
| Co-processing | Fats, oils, and, greases (FOG), from petroleum refining, biocrude | 2018 | 5% | Chevron, Phillips66, BP |
| Catalytic hydro thermolysis synthesized kerosene (CH-SK, or CHJ) | Fatty acids and fatty acid esters, lipids that come from plant and animal fats, oils and greases (FOGs) | 2020 | 50% | ARA |
| Hydro processed hydrocarbons (HH-SPK, or HC-HEFA) | Bio-derived hydrocarbons, directly from oils (triterpenes) produced by the *Botryococcus braunii* algae | 2020 | 10% | IHI Corporation |

ASTM Qualification Process (Heyne et al. 2021). Sustainability of these pathways depends upon the feedstock and way of production (CAAFI 2023; Csonka 2016).

One more difficulty in the use of biofuels is related to some less environmentally favorable components of conventional jet fuel (e.g. sulfur) that allows seals to swell in engines and prevent fuel leaks. New engines do not have this concern, and advanced biofuels have already been tested for performance at 100% in military aircraft. While biofuel production volumes remain low, it is expected that the blend limit will eventually increase to 100% (ATAG 2020).

In January 2023, the first flight in the Middle East and North Africa to be powered by 100% SAF was operated by Emirates in a Boeing 777-300ER. One engine used 100% SAF, the other engine used conventional jet fuel. The aircraft flew for more than 1 h. The SAF was obtained by blending two different SAFs in a ratio that mirrored the qualities of jet fuel: HEFA-SPK produced by Neste and HDO-SAK produced by Virent. The test flight demonstrates the compatibility of the specially blended SAF as a safe and reliable fuel source.

## 11.3 e-Fuels

E-Fuels are produced by combining hydrogen with carbon extracted from $CO_2$. Several processes are possible: power-to-gas in which gaseous fuels, such as methane, are produced; power-to-liquid, also known as synfuels, in which longer chained hydrocarbons from carbon dioxide are produced; and power-to-ammonia, in which ammonia is produced from nitrogen. The only inputs to the process of making them are fossil-free electricity, water, and recycled carbon dioxide, clearly in contrast to virgin fossil feedstock. e-Fuels, using renewable energy for fuel synthesis, are carbon-neutral with respect to GHG emission (Fig. 11.2).

The $CO_2$ source used as a feedstock for synthetic fuels can be existing point sources, that is, from existing industrial processes (cement production or power

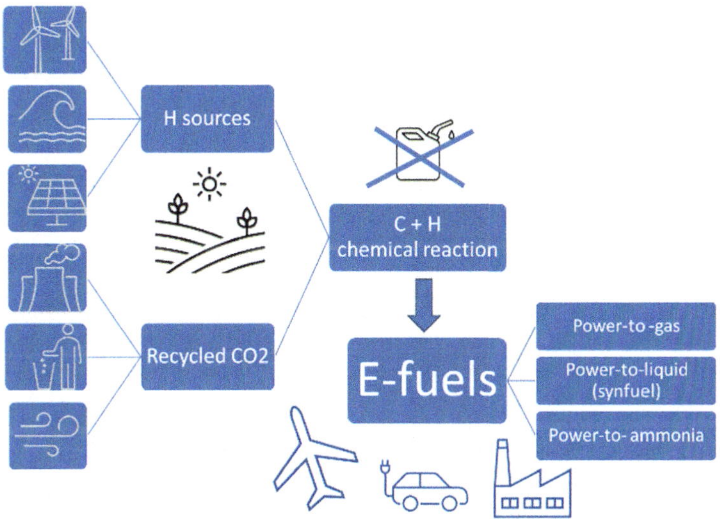


**Fig. 11.2** Schematic illustration of e-fuel production

generation) and biogenic sources (biomass, waste), or direct air capture, extracting $CO_2$ from the air. Direct air capture requires a lot more energy than extraction from a point source, which reduces the process efficiency. Indeed, the expected efficiencies for a Fischer–Tropsch pathway using high-temperature electrolysis are 47% in case of air capture and 64% in case of point source.

## 11.4 Comparison Between Advanced Biofuels and e-Fuels

Advanced biofuels cost 50% more than traditional jet fuel, the delivered cost of a feedstock accounting for the total costs of cultivation/plantation, harvesting, and other postharvest processing, storage, and transporting to the production plant. E-Fuels can cost 200% more, depending on the air capture method and the production of electricity.

In terms of infrastructure, since SAF are "drop-in" fuels, designed to be mixed with traditional fuels, no change or investment from airports is required as existing equipment can be used. E-Fuels have an increasing diffusion. An industrial consortium is planning Europe's first power-to-liquid plant in Norway: it will be producing renewable fuel from $CO_2$ and water using 100% renewable electricity.

E-Fuels will not be cost-competitive with conventional kerosene by 2030, and this will likely be the case for many years. As part of the "Fit for 55" package, the European Commission proposed to boost the uptake of SAF in air transport, with the binding blends percentages reported in Table 11.2, setting a sub-target to ensure that a certain amount of SAF used are synthetic fuels (EASA 2023).

## 11.5 Conclusions and Final Remarks

Only 4 years ago SAF accounted for just 0.1% of all jet fuel used worldwide. Currently, more than 450,000 flights have taken off using a mix of SAF and traditional fuels, more than 50 airlines have at least some experience with SAF, 7 technical pathways currently exist in creating types of SAF, more than 300 million liters of SAF were produced in 2022, and US$17 billion worth of SAF were in forward purchase agreements in 2022.

In the future, SAF will be the most used approach contributing toward sustainability targets before electric and hydrogen technologies become advanced. Electric and hybrid-electric technologies are currently the object of research projects such as

**Table 11.2** Binding blends proposed by the "Fit for 55" policy

| | 2025 | 2030 | 2035 | 2040 | 2045 | 2050 |
|---|---|---|---|---|---|---|
| Percentage of SAF used in air transport | 2% | 5% | 20% | 32% | 38% | 63% |
| Of which: sub-mandate for e-fuels | – | 0.7% | 5% | 8% | 11% | 28% |

the recently approved Horizon Europe Environmentally Friendly Aviation for All Classes of Aircraft (EFACA – https://cordis.europa.eu/project/id/101056866). When they are commercialized, the engines will be designed to be compatible to more promising biofuels such as e-fuels and liquid hydrogen, allowing a double environmental benefit. However, in the long term, SAFs will remain the only viable solution for sustainable long-distance aircraft.

## References


Ashcraft SW, Padron AS, Pascioni KA, Stout GW, Huff DL Jr, (2011) Review of Propulsion Technologies for N+3 Subsonic Vehicle Concepts. https://ntrs.nasa.gov/api/citations/20110022435/downloads/20110022435.pdf

CAAFI (2023) Fuel Qualification http://www.caafi.org/focus_areas/fuel_qualification.html

Chiaramonti D and Goumas T (2019), Impacts on industrial-scale market deployment of advanced biofuels and recycled carbon fuels from the EU Renewable Energy Directive II, Applied Energy, 2019, vol. 251, issue C, https://doi.org/10.1016/j.apenergy.2019

Comincini D (2018). Modular approach to hydrogen hybrid-electric aircraft design. Retrieved from http://hdl.handle.net/10589/143966

Csonka S (2016) CAAFI – CORE-JetFuel Cooperation Workshop. www.core-jetfuel.eu

EASA (2023) Fit for 55 and ReFuelEU Aviation. https://www.easa.europa.eu/en/light/topics/fit-55-and-refueleu-aviation

European Commission (2019) Communication from the commission to the European Parliament, the European Council, the council, the European Economic and Social Committee and the Committee of the regions: The European Green Deal, COM 640, https://eur-lex.europa.eu/legal-content/EN/TXT/?uri=COM:2019:640:FIN

Heyne J, Rauch B, Le Clercq P, Colket M (2021) Sustainable aviation fuel prescreening tools and procedures, Fuel, Volume 290, 120004, doi:https://doi.org/10.1016/j.fuel.2020.120004

IATA (2022) Net zero 2050: sustainable aviation fuels, www.iata.org/flynetzero

Misra A (2017). Technical Challenges and Barriers Affecting Turbo-electric and Hybrid Electric Aircraft Propulsion. Retrieved from https://ntrs.nasa.gov/archive/nasa/casi.ntrs.nasa.gov/20180004252.pdf

Seitz A, Schmitz O, Isikveren AT, Hornung M (2012). Electrically Powered Propulsion: Comparison and Contrast to Gas Turbines. Retrieved from https://www.dglr.de/publikationen/2012/281358.pdf

The European Parliament and the Council of the European Union (2018), Directive (EU) 2018/2001 of the European Parliament and of the Council on the promotion of the use of energy from renewable sources, Official Journal of the European Union. https://eur-lex.europa.eu/legal-content/EN/TXT/PDF/?uri=CELEX:32018L2001

Air Transport Action Group (ATAG) (2020) What is sustainable aviation fuel? https://aviationbenefits.org/faqs/what-is-sustainable-aviation-fuel/

# Chapter 12
# An Overview of the Possibilities, Current Status, and Limitations of Battery Technologies to Electrify Aviation

**Rosa Maria Arnaldo Valdés, Victor Fernando Gómez Comendador, Maria Zamarreño Suarez, Francisco Perez Moreno, and Raquel Delgado-Aguilera Jurado**

## 12.1 Aviation Electrification: A Path to Sustainable Air Travel

Aviation electrification is a game-changer, replacing fossil fuel propulsion with electric power for aircraft. Governments, research, manufacturers, and airlines are driving the shift. Initiatives like the EU's Clean Sky and NASA's X-57 Maxwell project accelerate battery tech integration.

Batteries are crucial for aviation electrification, powering electric aircraft, and benefiting from energy storage, power electronics, and electric motors. Their role includes propulsion, storage, hybrid-electric setups, safety, redundancy, and ground support. We will delve into battery tech possibilities, status, and limitations in aviation electrification in the following sections.

## 12.2 Aviation Battery Technologies: What They Need to Succeed

Aviation batteries play a vital role in safe, efficient flight. The key requirements for their success include

- *Energy density*: Refers to the amount of energy stored in a battery per unit mass or volume. Storing ample energy in a small space is necessary for long-distance flights.

R. M. A. Valdés (✉) · V. F. G. Comendador · M. Z. Suarez · F. P. Moreno · R. D.-A. Jurado
Universidad Politecnica de Madrid, Madrid, Spain
e-mail: rosamaria.arnaldo@upm.es

 
T. H. Karakoc et al. (eds.), *The Future of Electric Aviation and Artificial Intelligence*, Sustainable Aviation, https://doi.org/10.1007/978-3-031-62094-2_12

- *Power density*: Represents the rate at which a battery can deliver energy. Rapid energy delivery is necessary for quick demands like takeoff.
- *Cycle life*: Indicates the number of charge–discharge cycles a battery can undergo before its capacity degrades significantly. Aviation demands longevity with minimal capacity degradation.
- *Safety*: Robust features to prevent hazards.
- *Weight and space*: Lightweight, compact, and efficient designs.
- *Temperature tolerance*: Optimal performance across various conditions.
- *Fast charging*: Quick turnaround without compromising safety.
- *Environmental sustainability*: Nontoxic, recyclable, and eco-compliant materials.

Meeting these key requirements is critical to the successful integration of battery technologies into aviation and ensures that electric aircraft can operate reliably, efficiently, and safely.

## 12.3 Pay Attention to Performances

To analyze aircraft performance, we categorize them into regional, narrowbody, or widebody. Widebody aircraft carry more passengers on long journeys, regional aircraft serve short missions with fewer passengers, and narrowbody aircraft fall in between. Each category is characterized by average figures for empty weight fraction, wing aspect ratio, wing loading, drag coefficient, and mechanical efficiency data. With such characteristics, Monte Carlo simulations of the predetermined missions for each class of aircraft revealed that the typical regional aircraft would require 600 Wh/kg of battery density, narrowbodies 820 Wh/kg, and widebodies a whopping 1280 Wh/kg. Increasing specific energy with aircraft size is driven by their long-range use. Regional aircraft require around 500 Wh/kg, narrowbody needs 800 Wh/kg, and widebody requires 1700 Wh/kg to achieve respective thresholds. Even at these thresholds, electric widebody aircraft can only reach 16% of the current average range and passenger miles. Regional aircraft perform better at 2000 Wh/kg, exceeding current averages.

## 12.4 Battery Electrochemical Cell Technologies for Aviation Electrification

There could be different approaches for grouping and characterizing batteries. At a very high-level approach, batteries are commonly classified into two main groups: primary and secondary batteries. Primary batteries are nonrechargeable and not suitable for aviation due to disposal issues. Secondary batteries are rechargeable and can be classified into

- *Lithium-ion batteries*: popular in aviation due to high-energy density, though they have limitations.
- *Beyond lithium-ion batteries*:

  (a) *With insertion electrodes*: involves active ion insertion into an inactive host framework (e.g., aluminum, sodium).
  (b) *With conversion electrodes*: requires structural changes in electrodes during redox reactions, using metal anodes (e.g., lithium) combined with sulfur or oxygen.

- *All solid-state batteries*: use solid-state electrolytes to achieve higher energy densities.
- *Flow batteries*: store liquid electrolyte in external tanks, separating energy and power capabilities.

  These battery types offer various options for aviation applications.

### *12.4.1 Main Parameters and Relevant Characteristics of Different Battery Types for Aviation Applications*

Table 12.1 compares the key parameters and characteristics of various battery types. Table 12.2 summarizes the energy and power density potential of these batteries. Table 12.3 indicates the weight and travel distance based on energy density for each battery type. Finally, Table 12.4 summarizes the possibilities, current status, and limitations.

## 12.5 Conclusion

Aviation electrification aims to replace fossil fuel propulsion with electric power to reduce emissions. Batteries are vital, needing high-energy density, lightweight, efficiency, and temperature tolerance. This chapter addresses battery performance requirements, chemistry classification, and comparative analysis for aviation needs. The criteria include energy density, power density, cycle life, safety, size, weight, charging time, cost, weight, travel distance, safety, fast charging, life span, maintenance, raw materials, manufacturing, disposal, and TRL level.

**Table 12.1** Comparison of main parameters

| Battery type | Energy density | Power density | Cycle life | Safety | Size and weight | Charging time | Cost |
|---|---|---|---|---|---|---|---|
| Flow batteries | Medium to high | Low to medium | High | Safe | Large and heavy | Moderate | Moderate to high |
| All-solid batteries | High | Medium | High | High safety | Compact | Moderate | High |
| Beyond lithium-ion (insertion) | Medium to high | High | High | Moderate safety | Compact | Moderate | High |
| Beyond lithium-ion (conversion) | High | Medium | Moderate to high | Moderate safety | Compact | Moderate | High |
| Lithium-ion batteries | High | High | High | Moderate safety | Compact | Moderate | Moderate |

**Table 12.2** Potential energy density and power density

| | Lithium-ion batteries | Beyond lithium-ion (conversion electrodes) | Beyond lithium-ion (insertion electrodes) | All-solid batteries | Flow batteries |
|---|---|---|---|---|---|
| Energy density | 150 Wh/kg to 250 Wh/kg | 300 Wh/kg to 500 Wh/kg | 150 Wh/kg to 300 Wh/kg | 200 Wh/kg to 400 Wh/kg | 20 Wh/kg to 100 Wh/kg |
| Power density | 500 W/kg to 2000 W/kg | 100 W/kg to 500 W/kg | 200 W/kg to 1000 W/kg | 100 W/kg to 500 W/kg | 10 W/kg to 100 W/kg |

**Table 12.3** Weight and travel distance

| | Lithium-ion batteries | Beyond lithium-ion (conversion electrodes) | Beyond lithium-ion (insertion electrodes) | All-solid batteries | Flow batteries |
|---|---|---|---|---|---|
| Weight support | Good balance of energy and power for various aircraft sizes | High-energy density for larger aircraft with high power needs | Good balance of energy and power for different aircraft sizes | High-energy density, suitable for larger aircraft | Relatively lower energy density for smaller aircraft with lower power demands |
| Maximum travel distance | From hundreds to a few thousand kilometers | From hundreds to a few thousand kilometers | From hundreds to a few thousand kilometers | From hundreds to a few thousand kilometers | From tens to a few hundred kilometers |

**Table 12.4** Possibilities, current status, and limitations

| Insertion electrodes | | | |
|---|---|---|---|
| | Possibilities | Current status | Limitations |
| Sodium-ion (Na-ion) | Sodium resources more abundant; potential cost advantages. Potential high energy and longer life span | Research ongoing for better performance, stability, and safety. Not yet commercially available for aviation applications | Lower energy density, larger ions, and anode degradation concerns compared to Li-ion batteries |
| Aluminum-ion (Al-ion) | Aluminum batteries offer high energy density, safety, and cost benefits | Early development stage with technical challenges, including stability and dendrite formation | Facing difficulties in finding efficient electrolyte materials with high ionic conductivity and stability |
| Magnesium-ion (Mg-ion) | Magnesium batteries have higher energy density, improved safety, and cost advantages | In early development, researchers focus on materials for reversible magnesium-ion intercalation | Challenges for finding suitable electrolytes and electrode materials for magnesium-ion accommodation and high-performance cathodes |

(continued)

**Table 12.4** (continued)

| Insertion electrodes | | | |
|---|---|---|---|
| | Possibilities | Current status | Limitations |
| Zinc-ion (Zn-ion) | Zinc batteries: abundant, low-cost, and good energy storage | In early development, focusing on stability, energy density, and safety | Addressing dendrite formation and developing suitable electrolytes are challenges |
| *Conversion electrodes* | | | |
| Lithium-air (Li-air) | Ultra-high theoretical energy density for lightweight batteries | Early development with technical challenges, optimizing lithium-oxygen reaction, cathode, and electrolyte | Lithium peroxide formation, managing moisture, carbon dioxide, electrode, electrolyte stability, and lithium dendrite prevention |
| Lithium-selenium (Li-Se) | High energy density and safety potential compared to Li-S batteries, suitable for aviation | Early-stage development, focusing on performance optimization and addressing technical challenges | Challenges include selenium volume changes, reaction byproducts, and efficient charge transfer pathways |
| Flow batteries | Potential for higher energy storage, quick recharge, and safety for aviation | In early research and development stage with some promising proof-of-concept results | Lower energy and power density, larger size and weight, and infrastructure challenges for aviation implementation |

## References


Xin Shen, Xue-Qiang Zhang, Fei Ding, Jia-Qi Huang, Rui Xu, Xiang Chen, Chong Yan, Fang-Yuan Su, Cheng-Meng Chen, Xingjiang Liu, and Qiang Zhang. Advanced Electrode Materials in Lithium Batteries: Retrospect and Prospect. AAAS. Energy Material Advances. Volume 2021, Article ID 1205324, 15 pages. https://doi.org/10.34133/2021/1205324

Yulin Gao, Zhenghui Pan, Jianguo Sun, Zhaolin Liu, John Wang, High-Energy Batteries: Beyond Lithium-Ion and Their Long Road to Commercialisation. Nano-Micro Lett. (2022) 14:94. doi: https://doi.org/10.1007/s40820-022-00844-2

Grimaud, A. Batteries: Beyond intercalation and conversion. Nat Energy 2, 17003 (2017). doi: https://doi.org/10.1038/nenergy.2017.3

Wu, Y. (2014). Insertion Electrodes for Li Batteries. In: Kreysa, G., Ota, Ki., Savinell, R.F. (eds) Encyclopedia of Applied Electrochemistry. Springer, New York, NY. doi:https://doi.org/10.1007/978-1-4419-6996-5_309

Cary M. Hayner, Xin Zhao, and Harold H. Kung. Materials for Rechargeable Lithium-Ion Batteries Annual Review of Chemical and Biomolecular Engineering Vol. 3:445–471 (Volume publication date June 2012). https://doi.org/10.1146/annurev-chembioeng-062011-081024

Kuang, Chengwei; Zeng, Wen; Li, Yanqiong; A Review of Electrode for Rechargeable Magnesium Ion Batteries. Journal of Nanoscience and Nanotechnology, Volume 19, Number 1, January 2019, pp. 12–25(14). American Scientific Publishers. https://doi.org/10.1166/jnn.2019.16435

Elham Kamali-Heidari, Ata Kamyabi-Gol, Mahmoud Heydarzadeh sohi, Abolghasem Ataie Electrode Materials for Lithium Ion Batteries: A Review. Journal of Ultrafine Grained and

Nanostructured Materials. https://jufgnsm.ut.ac.ir. Vol. 51, No.1, June 2018, pp. 1–12. Print ISSN: 2423-6845 Online ISSN: 2423-6837. DOI: https://doi.org/10.22059/JUFGNSM.2018.01.01

Robert C. Massé, Chaofeng Liu, Yanwei Li, Liqiang Mai, Guozhong Cao. Energy storage through intercalation reactions: electrodes for rechargeable batteriesNational Science Review, Volume 4, Issue 1: 26–53 (2017). doi:https://doi.org/10.1093/nsr/nww093

Juyan Zhanga, Xuhui Yaob, Ravi K.Misrab, Jiong Caia, Yunlong Zhaob. Progress in electrolytes for beyond-lithium-ion batteries Journal of Materials. Science & Technology 2020,44(0): 237–257. doi:https://doi.org/10.1016/j.jmst.2020.01.017

Robert C. Massé, Evan Uchaker and Guozhong Cao Beyond Li-ion: electrode materials for sodium- and magnesium-ion batteries mater.scichina.com link.springer.com. https://doi.org/10.1007/s40843-015-0084-8 Sci China Mater 2015, 58: 715–766

# Chapter 13
# The Potential of a Proton Exchange Membrane Fuel Cell-Powered Light Aircraft Employing Cryogenic Hydrogen

J. Olsen

## 13.1 Introduction

Nearly all light aircraft today use AVGAS distilled from crude oil as their fuel source. As the cost of extracting the world's remaining supply of crude oil from the ground increases, we need to consider alternative fuels. In this chapter, we consider hydrogen as an alternative fuel source. The advantage of hydrogen is that its gravimetric specific energy is 2.75 times higher than AVGAS. Also, it is quiet and free from harmful emissions, provided that its energy is released by fuel cells. Its major disadvantage is its low density. At standard temperature and pressure, its density is 8400 times lower than AVGAS, making it difficult to physically store. In this work, we aim to see if can convert a Diamond DA40 piston engine aircraft operating on AVGAS into a fuel cell aircraft operating on cryogenic hydrogen.

## 13.2 Hydrogen Storage

The Diamond DA40 would require approximately 40 kg (390 N) of hydrogen to match the energy storage of AVGAS. If we pressurize hydrogen to a pressure of 70 MPa, its density is 18 times lower than AVGAS. As the weight of the tank required to store this pressurized hydrogen is around 14 times greater than the weight of the hydrogen itself, then a tank of 570 kg (5600 N) is required. This is clearly too much.

J. Olsen (✉)
The University of New South Wales, School of Mechanical and Manufacturing Engineering, Sydney, NSW, Australia
e-mail: j.olsen@unsw.edu.au



T. H. Karakoc et al. (eds.), *The Future of Electric Aviation and Artificial Intelligence*, Sustainable Aviation, https://doi.org/10.1007/978-3-031-62094-2_13

If we cryogenically cool hydrogen to a temperature of 20 K with its pressure at 1.2 bar, this saturated liquid hydrogen's (LH2) density is 10 times lower than AVGAS. A multilayered insulated tank is required to store the 390 N of LH2. Verstraete et al. (2010) suggest the following relationship: $W_T = \{(1 - \eta_g)/\eta_g\}$ $W_{LH2}$, where $W_T$ is the weight of the insulated tank, $W_{LH2}$ is the maximum weight of the LH2 stored, and $\eta_g = 0.675$ for a single multilayered insulated tank. The mass of tank required is 19.3 kg (189 N). As the density of the LH2 is 71.0 kg/m$^3$, then we require a tank whose volume is 570 liters. The weight of the tank should also include the weight of a heat exchanger, filling port, shut-off valve, etc., and so these items could increase the weight of the charged system to approximately 80 kg (785 N). This is far more acceptable. Note though that the fuel cell system is likely to have an efficiency almost double that of the engine it is replacing and so in practice the cryogenic hydrogen tank might be close to half the weight than that we have specified above.

## 13.3 Components of Hydrogen Fuel Cell Aircraft

The engine used in the Diamond DA40 fits within a rectangular prism, which is 52.3 cm high, 84.7 cm wide, and 84.5 cm long. It therefore occupies a volume of 0.374 m$^3$. Since it has a maximum power output of 134 kW, its volumetric power density is 358 kW/m$^3$ and its gravimetric power density is 1.145 kW/kg.

Based on the values found for the Toyota Mirai, the volumetric power density and gravimetric power density for the fuel cell stack are 3100 kW/m$^3$ and 2.04 kW/kg, respectively, for the electric motor are 6700 kW/m$^3$ and 13.0 kW/kg, respectively, for the boost controller are 126,800 kW/m$^3$ and 8.43 kW/kg, respectively, and for the battery are 5800 kW/m$^3$ and 0.45 kW/kg. These components can easily be placed within the space vacated by the engine. The cryogenic tank will need to be located behind the passenger compartment and not within the wings due to its bulky shape.

### *13.3.1 Fuel Cell Stack*

Our fuel cell stack consists of 350 individual proton exchange membrane (PEM) fuel cells stacked side by side. An individual fuel cell's voltage is a nonlinear function of its current density. The open-cell voltage is reduced as the current density is increased due to the effects of activation, ohmic, and concentration overvoltages (see Pukrushpan et al. 2004 for more detail). Note that this voltage reduction can be varied by quickly changing the pressure of the inlet air. The flow of protons across each individual fuel cell membrane is the product of both their current density and membrane area and is equal to the external electrical current.

Figure 13.1 shows the power produced by the fuel cell stack, which is the product of the cell voltage and the current that climbs to a maximum with current density

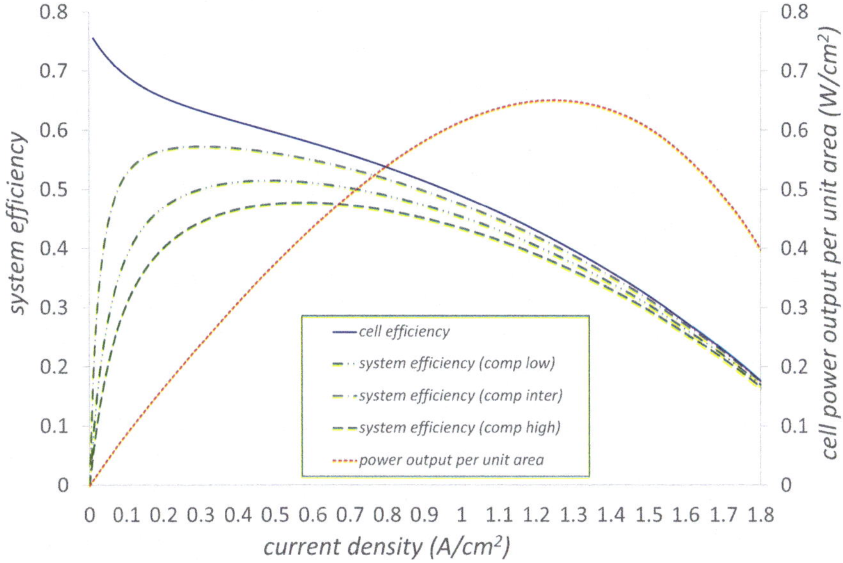


**Fig. 13.1** System efficiency, system efficiency with low compressor power, system efficiency with intermediate compressor power, and system efficiency with high compressor power and power output per unit area plotted against current density

before dropping off. It also shows the efficiency of the fuel cell stack, which is equal to the power output of the fuel cell stack divided by the sum of the power of the hydrogen that enters the fuel cell and the battery power that drives the air compressor and to a lesser extent the heating element that encourages the vaporization of LH2. The system efficiency is therefore reduced depending on how much power the battery provides. As a consequence, the system efficiency has a series of maximum values. These maxima decrease as the battery power requirement increases but they occur at higher current densities as the battery power requirement increases. In practice, we would like to run the fuel cell stack close to these maxima when the aircraft cruises near its design point. In doing so, it might be possible for the fuel cell system to operate when the aircraft cruises at an efficiency more than double that of an internal combustion engine running AVGAS.

Some typical numbers suggest that the fuel cell stack may provide maximum power at approximately 0.52 V/cell, and so given 350 cells, the voltage across the stack is $V_{\mathrm{fcs}} = 182$ V at a current density of 1.25 A/cm$^2$. If the membrane area is 640 cm$^2$, then the external electrical current is $I_{\mathrm{fcs}} = 800$ A. As the rotational speed of the motor is set by the propeller design at 251.3 rad/s, we will see that this current is too high and so we will require a boost controller (or DC/DC converter).

### 13.3.2 Boost Controller (DC/DC Converter)

Assuming minimal losses in the boost controller, the product of the output voltage and current of the fuel cell stack give the electrical power fed to the motor. However, the voltage and current need to be adjusted to enable the motor to spin at the required speed and enable the motor to match the load torque. A boost controller (or DC/DC converter) can increase the voltage applied to the motor through the following relationship: $V_m = V_{fcs}/(1 - D) = 535.3$ V, where $D = 0.66$ is the duty cycle. The current fed to the motor is $I_m = (V_{fcs}/V_m)\ I_{fcs} = (182/535.3)\ 800 = 272$ A (see Hughes and Drury 2013)

### 13.3.3 Axial Flux, Permanent Magnet, and Synchronous Machine

In an axial flux, permanent magnet, synchronous machine, the magnetic flux lines between the rotating permanent magnets and windings are parallel to the axis of the machine (see Gieras et al. (2008). These machines seem to have a higher gravimetric specific power output than radial flux machines. For our aircraft, we have designed a bespoke motor with $k = 2.0$ Nm/A and $R = 0.12\ \Omega$. For our motor, the back electromotive force $E$ is proportional to the rotational speed of the motor, that is, $E = k\omega$ and the load torque is proportional to the motor current, that is, $\tau = kI_m$. It is generally the case that the motor constant, designated by $k$, is equal, measured in Nm/A or V/1000 r/min (industry preferred). Our back electromotive force can be found from $E = V_m + I_mR = 502.7$ V forcing $\omega = 251.3$ rad/s (or 2400 r/min). The load torque the motor can accept is therefore $\tau = 544$ Nm. If we use a two-pole motor, the operating frequency will be 40 Hz; if we use a four-pole motor, the operating frequency will be 80 Hz, etc. Note that the power to the propeller $P_{prop} = \tau\ \omega = 136.7$ Nm with an efficiency of $\eta = 93.9\%$.

## 13.4 Conclusions

A light aircraft could be retrofitted with a fuel cell stack, electric motor, boost controller, larger lithium-ion battery, and a cryogenic hydrogen tank with a small weight penalty. This aircraft would have the same range capability as the AVGAS-powered aircraft it replaces, be quiet, and emission free.

## References

Gieras JF, Wand R, Kamper MJ, (2008) Axial Flux Permanent Magnet Brushless Machines, 2nd Edition, Springer.

Hughes A, Drury B (2013) Electric Motors and Drives, Fundamentals, Types, and Applications, 4th Edition, Newnes. https://doi.org/10.1016/j.rser.2019.02.022.

Pukrushpan JT, Peng H, Stefanopoulou AG, (2004) Control-Oriented Modelling and Analysis for Automotive Fuel Cell Systems, Transactions of the ASME, Vol. 126, pp14–25.

Verstraete D, Hendrick O, Pilidis P, Ramsden K (2010) Hydrogen Fuel Tanks for Subsonic Transport Aircraft. International Journal of Hydrogen Energy. https://doi.org/10.1016/j.ijhydene.2010.06.060

# Chapter 14
# Fault-Tolerant Estimation of UAV Dynamics in the Presence of Sensor/Actuator Faults

**Chingiz Hajiyev**

## Nomenclature

| | |
|---|---|
| AKF | Adaptive Kalman filter |
| EKF | Extended Kalman filter |
| FDI | Fault detection and isolation |
| IAE | Innovation-based adaptive estimation |
| MMAE | Multiple-model-based adaptive estimation |
| NQI | Quadratic innovation function |
| RAE | Residual-based adaptive estimation |
| RAKF | Robust adaptive Kalman filter |
| TSKF | Two-stage Kalman filter |
| UAV | Unmanned aerial vehicle |

## 14.1 Introduction

In aerospace control systems, sensor/actuator failures can deteriorate the closed-loop performance considerably. Such failures must be diagnosed as quickly as possible and actively accommodated to arrest the performance degradation. Sensor/actuator fault detection and isolation (FDI) can be achieved by employing model-based FDI techniques.

Many model-based FDI techniques have been developed to detect and identify sensor/actuator faults by using the Kalman filtering (KF) approach (Mohamed and Schwarz 1999; Hajiyev and Caliskan 2005; Caliskan and Hacizade 2014). The KF

C. Hajiyev (✉)
Istanbul Technical University, Faculty of Aeronautics and Astronautics, Istanbul, Turkiye
e-mail: cingiz@itu.edu.tr



T. H. Karakoc et al. (eds.), *The Future of Electric Aviation and Artificial Intelligence*, Sustainable Aviation, https://doi.org/10.1007/978-3-031-62094-2_14

approach to the state estimation is quite sensitive to any malfunctions. If the real-time condition of the system does not correspond to the models used in the synthesis of the filter, then these changes resulting from some possible failures at the sensors or actuators significantly decrease the effectiveness of the estimation system. In this case, the adaptive Kalman filters (AKFs) can be used to recover the possible malfunctions.

By using different techniques, the KF can be made adaptive and insensitive to a priori sensor or actuator failures. Three basic approaches to the adaptive Kalman filtering problem are the multiple-model-based adaptive estimation (MMAE) (Maybeck 1999), innovation-based adaptive estimation (IAE) (Mohamed and Schwarz 1999), and residual-based adaptive estimation (RAE) (Mohamed and Schwarz 1999; Wang et al. 1999).

Another concept is to scale the noise covariance matrix by multiplying with a time-dependent variable (Hu et al. 2003; Hajiyev 2007; Hajiyev and Soken 2012). An adaptive Kalman filter is used to tune the measurement and process noise covariance matrices $R$ and $Q$, respectively. When the process noise covariance matrix is scaled ($Q$-adaptation), it means that there is an actuator fault to be compensated. Otherwise, when the measurement noise covariance matrix is scaled ($R$ adaptation), it is a sensor fault, which is dealt with.

In this study, the combined fault isolation and $R$- and $Q$-adaptation procedures are solved via one robust adaptive Kalman filter (RAKF) incorporating the $R$- and $Q$-adaptation filters. After the detection of faults, it is proposed to run simultaneously both $R$-adaptive and $Q$-adaptive Kalman filters to distinguish the sensor and actuator faults. For the fault isolation purpose, estimation performances of the $R$- and $Q$-adaptation procedures are compared. As the estimation performance criterion index, it is proposed to select the sum of quadratic differences between estimation and extrapolation values of investigated filters.

## 14.2 Robust Adaptive Kalman Filtering

For the investigated estimation problem, we preferred to work with the combined model for the longitudinal and lateral equations of motion. Hence, the state vector to be estimated is formed of $n = 9$ states as

$$x = [\Delta u \;\; \Delta w \;\; \Delta q \;\; \Delta \theta \;\; \Delta h \;\; \Delta \beta \;\; \Delta p \;\; \Delta r \;\; \Delta \phi]^T, \tag{14.1}$$

while the control input vector is

$$u = [\Delta \delta_e \quad \Delta \delta_T \quad \Delta \delta_a \quad \Delta \delta_r]^T. \tag{14.2}$$

Here, $\Delta u$, $\Delta v$, $\Delta w$ are the velocities along $x$, $y$, and $z$ directions in the body frame; $\Delta p$, $\Delta q$, $\Delta r$ are the angular rates around $x$, $y$, and $z$ axes; $\Delta \theta$ is the pitch angle; $\Delta \phi$ is the roll angle; $\Delta\beta$ is the sideslip angle; $\Delta h$ is the height; $\Delta\delta_e$, $\Delta\delta_a$, and $\Delta\delta_r$ are the elevator, aileron, and rudder deflections; and $\Delta\delta_T$ is the change in the thrust. All of these states are measured directly so that the measurement matrix that relates the states with the measurements is a $9 \times 9$ identity matrix.

At that rate, an optimal KF algorithm for the unmanned aerial vehicle (UAV) state estimation can be easily built and accurate estimations can be obtained if there is no fault in the system. On the other hand, as mentioned above, if a sensor or actuator fault is in question, then a RAKF algorithm should be used instead.

The RAKF algorithm that we propose is composed of the *R*- and *Q*-adaptation procedures. Depending on the type of fault (either it is an actuator fault that causes the changes in the control derivatives or a sensor fault that causes the abnormal measurements, increase in the background noise, etc.), the filter performs an adaptation procedure on the process noise covariance matrix, *Q*, or measurement noise covariance matrix, *R*. A robust method with *R*-adaptation is called robust Kalman filter (RKF), and an adaptive method with *Q*-adaptation is called adaptive Kalman filter (AKF) in this chapter. The RKF and AKF algorithms can be found in Hajiyev and Soken 2013.

## 14.3 The Sensor/Actuator Fault Detection and Isolation

The fault detection process is applied by checking a statistical function at each estimation step. The following two hypotheses may be introduced: $\gamma_0$, the system is normally operating; $\gamma_1$, there is a malfunction in the system. In order to detect the faults, the statistical function [normalized quadratic innovation function (NQI)] can be used:

$$\mathrm{NQI}(k) = \Delta^T(k)\left[\mathrm{HP}(k/k-1)H^T + R(k)\right]^{-1}\Delta(k) \tag{14.3}$$

where $\Delta(k)$ is the innovation sequence, *H* is the measurement matrix, $P(k/k-1)$ is the covariance matrix of the extrapolation error, and $R(k)$ is the covariance matrix of the measurement noise.

This statistical function has $\chi^2$ distribution with *N* degree of freedom, where *N* is the dimension of the innovation vector. By selecting the level of significance, $\alpha$, the threshold value, $\chi^2_{\alpha,N}$, can be found. Hence, when the hypothesis $\gamma_1$ is correct, the statistical value of NQI($k$) will be greater than the threshold value $\chi^2_{\alpha,N}$, that is,

$$\begin{aligned} \gamma_0 &: \mathrm{NQI}(k) \le \chi^2_{\alpha,N} \quad \forall k \\ \gamma_1 &: \mathrm{NQI}(k) > \chi^2_{\alpha,N} \quad \exists k. \end{aligned} \tag{14.4}$$

After detecting the fault in the system, the key point is detecting the type of the fault (either a sensor or actuator fault). Then, the appropriate adaptation (the *R*- or

$Q$-adaptation) may be applied. In this study, the combined fault isolation and $R$- and $Q$-adaptation procedures are solved via one RAKF incorporating the $R$- and $Q$-adaptation filters. The proposed approach does not require the additional fault isolation filter.

When a KF is used, the decision statistic (14.3) changes regardless of whether the fault is in the sensors or the actuators. After the detection of faults, it is proposed to run simultaneously both $R$-adaptive and $Q$-adaptive Kalman filters to distinguish the sensor and actuator faults. For the fault isolation purpose, comparison of the estimation performances of the $R$- and $Q$-adaptation procedures is proposed.

As an indicator of the estimation quality criterion, it is proposed to choose the quadratic differences between the estimated $\widehat{x}(k/k)$ and extrapolation $\widehat{x}(k/k-1)$ values of the studied filters after the detection of a fault at the iteration $k = \tau$

$$\varepsilon(k) = [\widehat{x}(k/k) - \widehat{x}(k/k-1)]^T [\widehat{x}(k/k) - \widehat{x}(k/k-1)], k \geq \tau \qquad (14.5)$$

In the case of sensor fault, the estimation performance of the $R$-adaptation KF will be better (the values of criterion (14.5) for RKF will be lower); and in the case of actuator fault, the $Q$-adaptation KF will give better estimation results (the values of criterion (14.5) for the RKF will be larger). If the actuator fault occurs, the $Q$-adaptation is implemented; otherwise, the procedure continues with the $R$-adaptation.

## 14.4 Simulations

The proposed combined sensor/actuator fault isolation and $R$- and $Q$-adaptation filters are applied to the model of dynamics of an experimental platform ZAGI UAV (Matthews 2006). The simulations are realized in 1000 steps for a period of 100 s with 0.1 s of sampling time. The algorithms are tested for two different measurement malfunction scenarios: continuous bias at measurements and measurement noise increment. In case of sensor/actuator fault, the simulations are carried out for the RKF and AKF to compare the results of state estimation and performance criterion (14.5).

### *14.4.1 Continuous Bias at Measurements*

Continuous bias term is simulated by adding a constant term to the measurements of pitch angle gyro in the interval $300 \leq k \leq 600$. As the estimation performance criterion, it is proposed to select the sum of quadratic differences between estimation and extrapolation values of the investigated filters for $m$ realizations after the detection of faults at the $k = \tau$iteration,

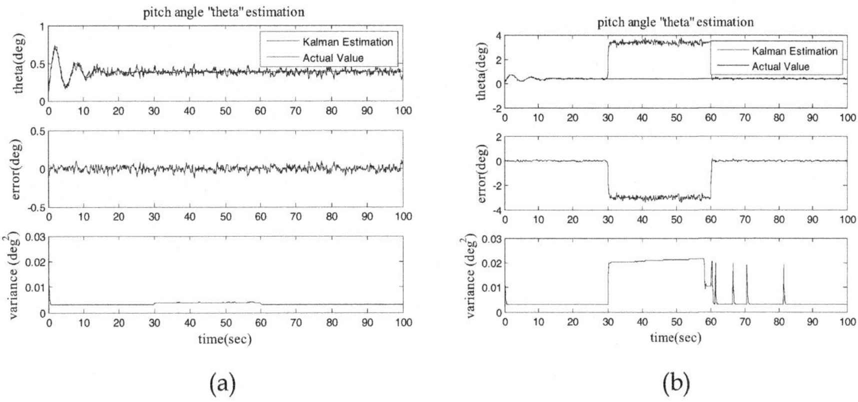


**Fig. 14.1** Estimation results for the pitch angle by the RKF (**a**) and AKF (**b**) in the presence of pitch angle gyro bias

$$\varepsilon^* = \sum_{k=\tau}^{\tau+m} [\widehat{x}(k/k) - \widehat{x}(k/k-1)]^T [\widehat{x}(k/k) - \widehat{x}(k/k-1)], k \geq \tau \tag{14.6}$$

The performance criteria (14.6) allow us to compare the RKF and AKF performances easily and decrease the false decision probability. Using the number of realizations $m = 10$, the performance criterion values (14.6) for RKF and AKF are obtained, respectively, as $\varepsilon^*_{\text{RKF}} = 0,5111$ and $\varepsilon^*_{\text{AKF}} = 106,8903$. As shown, $\varepsilon^*_{\text{RKF}} << \varepsilon^*_{\text{AKF}}$, this inequality corresponds to the sensor fault in the system and the estimation procedure is recommended to continue with the RKF.

The pitch angle estimation results of RKF and AKF in the presence of pitch angle gyro bias are given in Fig. 14.1a, b. Figure 14.1 consists of three subfigures. The top gives the RKF or AKF state estimation results and the actual values by comparison. The middle shows the error of the estimation process based on the actual state values of the UAV. The last and the lowest indicates the variance of the estimation that is obtained as the diagonal elements of the KF estimation error covariance matrix.

As is clear from Fig. 14.1a, b, the RKF gives superior estimation results than the AKF in the case of pitch angle gyro bias. The RKF secures accurate estimation results for the whole process while the AKF fails in pitch angle gyro fault conditions.

### *14.4.2 Measurement Noise Increment*

In the second measurement malfunction scenario, measurement noise increment fault is simulated by multiplying the standard deviation of the noise of the pitch angle gyro measurements with a constant term in the interval $300 \leq k \leq 600$. For $m = 10$, the performance criterion values (14.6) for RKF and AKF are obtained, respectively, as $\varepsilon^*_{\text{RKF}} = 1,5840$ and $\varepsilon^*_{\text{AKF}} = 13,4121$. As is clear $\varepsilon^*_{\text{RKF}} < \varepsilon^*_{\text{AKF}}$, and

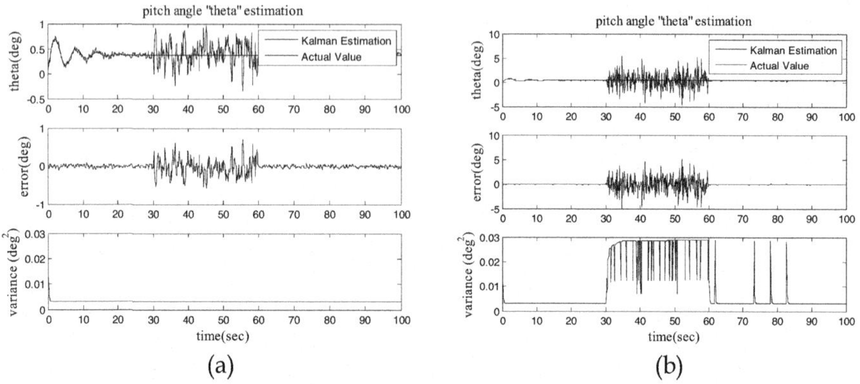


**Fig. 14.2** Estimation results for the pitch angle by the RKF (**a**) and AKF (**b**) in the presence of pitch angle gyro noise increment

the sensor fault in the system is detected. In this case, the estimation procedure should be continued with the RKF. The pitch angle estimation results of RKF and AKF in the presence of pitch angle gyro noise increment are presented in Fig. 14.2a, b.

### 14.4.3 Actuator Fault

In the third malfunction scenario, the actuator fault case is considered. In simulations, the actuator fault is simulated by multiplying the first column elements of the UAV control distribution matrix with a constant term 0.001 in between the 30th and 60th seconds. This fault corresponds to the loss of effectiveness of elevator.

Using the number of realizations $m = 10$, the performance criterion values (14.6) for RKF and AKF are found, respectively, as $\varepsilon^*_{\text{RKF}} = 5,3180$ and $\varepsilon^*_{\text{AKF}} = 3.6284$. As is clear $\varepsilon^*_{\text{AKF}} < \varepsilon^*_{\text{RKF}}$, this inequality corresponds to the actuator fault in the system and the estimation procedure is recommended to continue with the AKF.

The pitch angle estimation results of RKF and AKF in the presence of elevator loss of effectiveness are presented in Fig. 14.3a, b. As seen in Fig. 14.3a, b, the AKF gives superior estimation results than the RKF in the presence of actuator fault.

## 14.5 Conclusion

In this study, two opposite strategies are proposed in the occasion of sensor/actuator faults in system: first, the adaptive strategy, which is insensitive to the actuator failures ($Q$-adaptation strategy); and second, the robust strategy, capable of being insensitive to measurement errors ($R$-adaptation strategy). A novel adaptation procedure is proposed with an approach of making a choice among the $R$-adaptation and

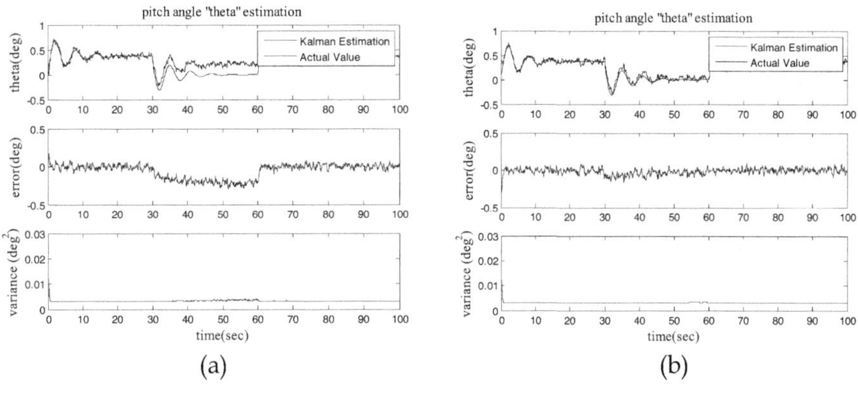


**Fig. 14.3** Estimation results for the pitch angle by the RKF (**a**) and AKF (**b**) in the presence of actuator fault

$Q$-adaptation strategies. In the novel method, after the detection of faults, it is proposed to run simultaneously both $R$-adaptive and $Q$-adaptive Kalman filters and compare their estimation performances to distinguish the sensor and actuator faults. As the estimation performance criterion index, the sum of square of the differences between estimation and extrapolation values of investigated filters is used.

# References


Caliskan F, Hacizade C (2014) Sensor and Actuator FDI Applied to an UAV Dynamic Model, in: 19th World Congress The International Federation of Automatic Control, Cape Town, South Africa: 12220–12225.

Hajiyev C (2007) Adaptive Filtration Algorithm with the Filter Gain Correction Applied to Integrated INS/Radar Altimeter. Proc. of the Institution of Mechanical Engineers, Part G: Journal of Aerospace Engineering 221: 847–885.

Hajiyev C, Caliskan F (2005) Sensor and control surface/actuator failure detection and isolation applied to F-16 flight dynamics. Aircraft Engineering and Aerospace Technology 77: 152–160.

Hajiyev C, Soken HE (2012) Robust Estimation of UAV Dynamics in the Presence of Measurement Faults. Journal of Aerospace Engineering 25(1): 80–89.

Hajiyev C, Soken HE (2013) Robust Adaptive Kalman Filter for Estimation of UAV Dynamics in the Presence of Sensor/Actuator Faults. Aerospace Science and Technology 28(1): 376–383.

Hu C, Chen W, Chen Y, and Liu D (2003) Adaptive Kalman Filtering for Vehicle Navigation. Journal of Global Positioning Systems 2: 42–47.

Matthews JS (2006) Adaptive Control of Micro Air Vehicles, M.Sc. Thesis, Department of Electrical and Computer Engineering, Brigham Young University, Utah, USA.

Maybeck PS (1999) Multiple Model Adaptive Algorithms for Detecting and Compensating Sensor and Actuator/Surface Failures in Aircraft Flight Control Systems. International Journal of Robust and Nonlinear Control 9(14): 1051–1070.

Mohamed AH, Schwarz KP (1999) Adaptive Kalman filter for INS/GPS. Journal of Geodesy 73: 193–203.

Wang J, Stewart MP, Tsakiri M (1999) Adaptive Kalman Filtering for Integration of GPS with GLONASS and INS, in: Proceeding of the 1999 International Geodesy Symposia (IUGG 99), Birmingham, UK.

# Chapter 15
# Hydrogen Storage in a Commuter Aircraft: Combining Classical Engineering Design Process with Model-Based System Engineering for CFRP Pressure Vessel Integration

**S. Spitzer, M. Peciak, F. Dexl, F. Schmidt, H. Antonowitz, A. Langkamp, W. Skarka, J. Markmiller, and M. Gude**

## Nomenclature

| | |
|---|---|
| CFRP | Carbon fiber-reinforced polymer |
| MBD | Model-based design |
| MBSE | Model based system engineering |

## 15.1 Introduction

Due to its low volumetric energy density, hydrogen must be stored under high-pressure or cryogenic conditions. The integration of hydrogen tanks into aircraft therefore raises new challenges compared to conventional tanks for fossil fuels. At

S. Spitzer (✉) · F. Schmidt · A. Langkamp · M. Gude
Institute of Lightweight Engineering and Polymer Technology, Technische Universität Dresden (TUD), Dresden, Germany
e-mail: sebastian.spitzer@tu-dresden.de; florian.schmidt@tu-dresden.de

M. Peciak · W. Skarka
Department of Fundamentals of Machinery Design, Silesian University of Technology, Gliwice, Poland
e-mail: magdalena.peciak@polsl.pl

F. Dexl · J. Markmiller
Institute of Aerospace Engineering, TUD, Dresden, Germany
e-mail: florian.dexl@tu-dresden.de

H. Antonowitz
Leichtbau-Zentrum Sachsen GmbH, Dresden, Germany
e-mail: henrik-antonowitz@lzs-dd.de



T. H. Karakoc et al. (eds.), *The Future of Electric Aviation and Artificial Intelligence*, Sustainable Aviation, https://doi.org/10.1007/978-3-031-62094-2_15

the same time, designing an aircraft with an alternative propulsion system brings with it the lack of empirical data on the new propulsion system. This makes it difficult to answer the question of whether the aircraft design under development will meet the performance requirements. Model-based design (MBD) can be used to reduce errors in the early stages of design and minimize costs and time. This approach places models at the center of the design process from the outset. MBD is particularly useful when dealing with dynamic systems such as aircraft. An additional advantage is that the models developed can be used in other projects, thus speeding up the design phase (Kelemenová et al. 2013; Peciak and Skarka 2022). For such a model-based approach, knowledge of the subsystems is crucial. In this case, the hydrogen storage subsystem is considered as a white box and the necessary information is generated through the classical engineering design process. It is expected that commuter up to short-range aircraft will be the first aircraft categories in which hydrogen-powered aircraft will enter service (Gao et al. 2022). Therefore, a commuter aircraft is considered the baseline.

## 15.2 Integration of Hydrogen Tank in the Aircraft Wing

The integration of hydrogen tanks into the wing seems (cf. Fig. 15.1) an attractive solution to avoid aerodynamical disadvantages compared to underwing pods.

However, as aircraft wings are significantly deformed during flight, design of integrated tanks must account for both the internal pressure loading and the bending of the wing. The latter results in either assuming the aerodynamic wing loads partially bore by the tank as external loads, or by defining a suitable design space in the deformed wing that allows a tank bearing without external loads.

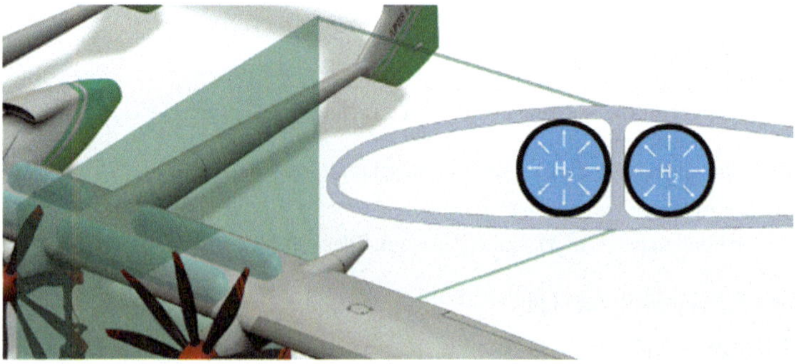


**Fig. 15.1** Concept of high-pressure hydrogen tanks integrated in the cross section of an aircraft wing

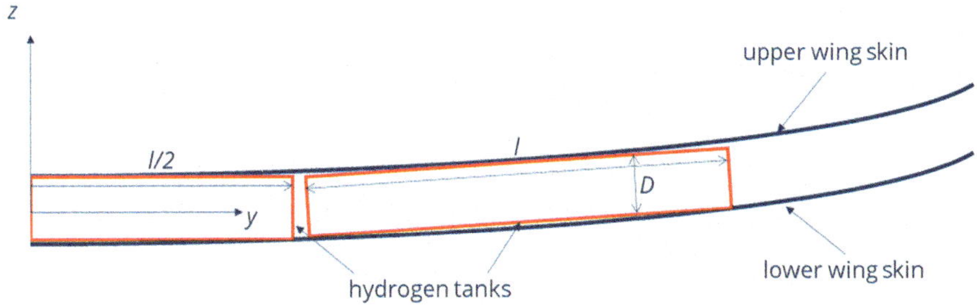


**Fig. 15.2** Schematic illustration of non-load-bearing tanks in aircraft wing bent due to aerodynamic loads

### *15.2.1 Pressure Vessel Design*

Based on current technologies with a sufficient technology readiness level, non-load-bearing tanks are chosen as a starting point. To determine the allowed design space, the wing bending is calculated first (cf. Fig. 15.2).

The tanks are sized according to Barlow's formula and the optimal length $l$ and outer diameter $D$ of the spanwise-segmented tank are determined. For the chosen reference aircraft and a working pressure $p = 700$ bar, an optimum of two spanwise-segmented tanks per half wing with $l = 2.53$ m and $D = 202$ mm results.

### *15.2.2 Virtual and Physical Demonstration of the Manufacturability*

Next, the tank is modeled using CADWIND (cf. Fig. 15.3 (left)) in order to develop a suitable fiber deposition process used for manufacturing a demonstrator (cf. Fig. 15.3 (right)).

## 15.3 Development of a Simulation Model for Aircraft Performance Evaluation

Using the MBD approach, a system-level simulation model of the aircraft has been developed to assess the energy and power requirements during flight.

It is possible to size the propulsion components accordingly, check the amount of hydrogen required stored in the tanks, and thus verify and optimize the flight parameters. The model is based on the forces acting on the aircraft and consists of four main subsystems (cf. Fig. 15.4): flight control module, environment, airframe

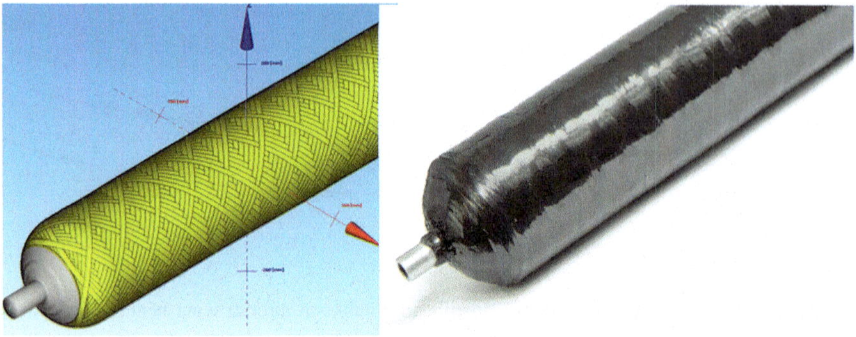

**Fig. 15.3** Winding core covered with fibers (yellow) in the software CADWIND (left) and manufactured CFRP vessel (right)

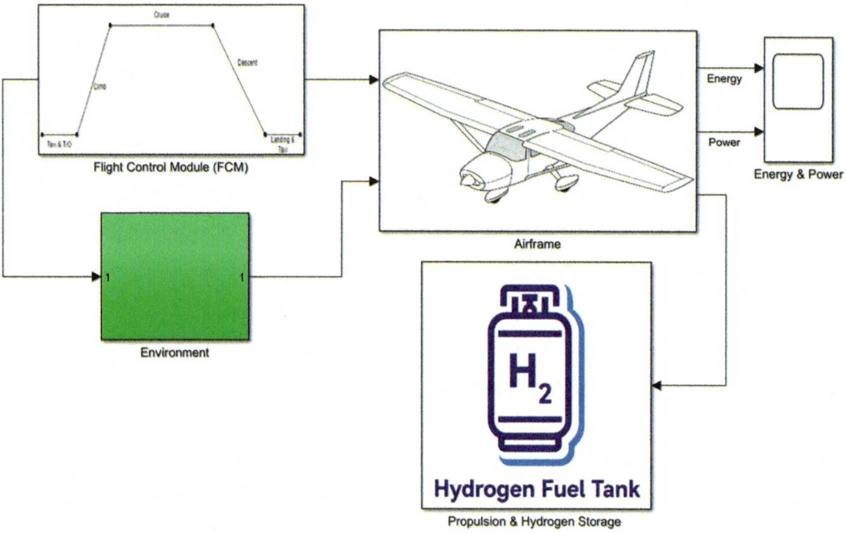


**Fig. 15.4** Simulation model developed with MATLAB/Simulink software to assess the energy and power demand of the analyzed aircraft

(in Fig. 15.4), and propulsion with hydrogen storage. Currently, the model is still in development to simulate the aircraft's behavior due to the placement of drive components, including hydrogen tanks, and analyze the propulsion parameters. This will provide a more holistic view of the propulsion system and the aircraft as a whole.

## 15.4 Conclusion

Model-based design allows the description of complex overall systems and can be used to model new aircraft with alternative fuel systems. Due to the low volumetric storage density of hydrogen, the integration of a sufficient amount of the energy carrier into the aircraft structure is a major challenge. There is limited empirical information available, so the subsystems of an MBSE model must be converted from a subsystem described by its boundaries (black box) to a subsystem where the inside is explored and described (white box). The necessary information can be generated by applying classical and established interactive engineering design processes. The disciplines of design, layout, and manufacturing are applied to the development of CFRP high-pressure tanks. In a combined virtual-physical approach, tanks have been developed that can be integrated into the center wing of a commuter aircraft. In the next step, the tanks developed in this way will have to be integrated into the center wing by means of CAD modeling and the corresponding mountings will have to be worked out. This will allow the input data for the model-based design to be derived in a well-founded manner, enabling the aircraft to be designed and its range to be estimated.

**Acknowledgments** The research was done in the project "System-integrated H2 pressure tank assemblies for general aviation and commuter aircraft" and financed by the European Regional Development Funding (EFRE) in Saxony: Technology Funding, Research and Development with the project number 100558500, funded by the Sächsische Aufbaubank (SAB). Part of the work was supported by DAAD: Research Grants - Bi-nationally Supervised Doctoral Degrees/Cotutelle, 2021/22 (57552338).

## References

Gao Y, Jausseme C, Huang Z, Yang T (2022) Hydrogen-powered aircraft: Hydrogen-electric hybrid propulsion for aviation. IEEE Electrification Magazine 10(2):17–26. https://doi.org/10.1109/MELE.2022.3165725

Kelemenová T, Kelemen M, Miková Ľ, Maxim V et al. (2013): Model Based Design and HIL Simulations. American Journal of Mechanical Engineering 1(7):276–281. http://pubs.sciepub.com/ajme/1/7/25/

Peciak, M, Skarka W (2022): Assessment of the Potential of Electric Propulsion for General Aviation Using Model-Based System Engineering (MBSE) Methodology. Aerospace 9(2):74. https://doi.org/10.3390/aerospace9020074

# Chapter 16
# Review of Liquid Hydrogen Tanks' Design Principles for Short- and Medium-Range Civil Flights

**Abdullah Bamoshmoosh and Gianluca Valenti**

## 16.1 Introduction

The aviation sector plays a pivotal role in global transportation, facilitating economic growth and fostering global connectivity. However, its contribution to climate change through the emission of greenhouse gases (GHGs), particularly carbon dioxide, has raised concerns regarding the sector's sustainability. As the demand for air travel continues to soar, it becomes imperative to decarbonize the aviation industry to mitigate its environmental impact and align with global climate goals. Decarbonizing the aviation sector is crucial for several reasons. Firstly, aviation emissions contribute significantly to global GHG emissions. According to the International Civil Aviation Organization, aviation accounts for about 2–3% of global $CO_2$ emissions (International Civil Aviation Organization 2019). While this share may appear relatively small, the unique characteristics of aviation emissions, such as the release of pollutants at high altitudes, result in a disproportionately higher climate impact. Additionally, the Intergovernmental Panel on Climate Change (IPCC) projects that aviation emissions could triple by 2050 if no substantial mitigation actions are taken (Intergovernamental Panel on Climate Change 2018). Thus, decarbonizing the aviation sector is essential to achieving the goals outlined in international agreements, such as the Paris Agreement, and limiting global warming to well below 2 °C.

As the aviation sector strives to decarbonize and reduce its environmental impact, the search for sustainable fuel alternatives has gained significant attention. Among the emerging options, liquid hydrogen has emerged as a promising solution that offers the potential to revolutionize the aviation industry. Liquid hydrogen has attracted interest as an enabler to achieving a sustainable aviation future because

A. Bamoshmoosh (✉) · G. Valenti
Department of Energy, Politecnico di Milano, Milan, Italy
e-mail: abdullah.bamoshmoosh@polimi.it

 
T. H. Karakoc et al. (eds.), *The Future of Electric Aviation and Artificial Intelligence*, Sustainable Aviation, https://doi.org/10.1007/978-3-031-62094-2_16

of its high-energy density and zero-carbon emissions. As a clean fuel, its combustion produces only water vapor, completely eliminating carbon dioxide typically associated with conventional fossil fuels (Abe et al. 2019). Furthermore, advancements in hydrogen production, storage, and distribution technologies have accelerated the potential for scalable and sustainable hydrogen infrastructure for aviation applications (Baroutaji et al. 2019). However, challenges such as hydrogen availability, infrastructure development, and safety considerations still need to be addressed to enable the widespread adoption of liquid hydrogen in the aviation sector. Nevertheless, the immense potential of liquid hydrogen as a zero-emission fuel presents an exciting opportunity for the aviation industry to embrace a sustainable and environmentally responsible path forward. This chapter aims at giving a brief overview of the specific design strategies and concepts behind the storage of liquid hydrogen for aircraft.

## 16.2 Cryogenic Tank Design

Cryogenic fluids are stored in tanks that are specifically designed to minimize heat transfer between the fluid and the surrounding environment. This is because heat leaks lead to an increase in pressure of the tank that can lead to structural damage. In the case of cryogenic liquids, this is of major relevance. This is because cryogenic liquids are generally stored in saturated conditions. The increase in pressure with temperature thus follows the saturation curve, resulting in exponential growth of pressure, increasing the risk of failure.

In general, four main strategies to insulate cryogenic tanks can be identified (Timmerhaus and Flynn 1989): vacuum insulation, multilayer insulation, powder insulation, and foam insulation. All of these strategies imply the use of an inner storage and an outer vessel. These strategies are not mutually exclusive. In fact, generally speaking, they target different heat transfer mechanisms and can be combined. Vacuum insulation eliminates convection heat transfer as it greatly reduces the number of molecules surrounding the inner vessel. As the density of the residual gas is decreased, conductive heat transfer between the inner and outlet vessel is also greatly reduced (Poling et al. 2001). Multilayer insulation consists of up to 80 layers of low-emittance sheets that are placed in order to minimize conductive heat transfer. This is obtained either by inserting low-conductivity spacers or minimizing the contact surface between the sheets. Powder insulation consists of filling the cavity between the inner and outer vessel with a low-conductivity, low-density powder. Because of the small size of the particles and the void between them, both heat conduction and convection are disrupted. Foam insulation consists of the use of expanded organic solids, mainly polyurethane, to cover the vessel. Because of the low density of the foam, heat conduction in the foam is greatly reduced. On the other hand, the continuous cavities in the foam make this strategy less effective against heat convection.

# Chapter 17
# New Supersonic Aircraft Emission and Air Pollution Assessment at Airport Operational Scenario

**K. Synylo, O. Zaporozhets, A. Krupko, and T. Reklewski**

## 17.1 Introduction

The Concorde Environment Impact Assessment (EIA) study at Dulles International Airport (IAD) demonstrated that the contribution of aircraft emissions to air quality inside the airport occurs during two operational modes: the queuing and low-speed ground run prior to takeoff of the aircraft [1]. Such single-event measurements were used to define the concentrations from aircraft emission and, in particular, the Concorde influence area referred to in Fig. 17.1. The measurement procedure was to place three monitoring stations downwind of the Concorde movement path (including the stage of its start, taxi, and takeoff) and measure the dispersion of the emission plume as it was transported by the wind over these stations. The concept of the *area of Concorde influence* has been introduced to its emissions blend (CO, HC, and NOx) and the lowest concentrations normally detectable above the background.

K. Synylo (✉) · A. Krupko
National Aviation University, Kyiv, Ukraine

O. Zaporozhets
Łukasiewicz Research Network – Institute of Aviation, Warsaw, Poland
e-mail: Oleksandr.Zaporozhets@ilot.lukasiewicz.gov.pl

T. Reklewski
National Fund for Environmental Protection and Water Management, Warszawa, Poland
e-mail: Tadeusz.Reklewski@nfosigw.gov.pl

 
T. H. Karakoc et al. (eds.), *The Future of Electric Aviation and Artificial Intelligence*, Sustainable Aviation, https://doi.org/10.1007/978-3-031-62094-2_17

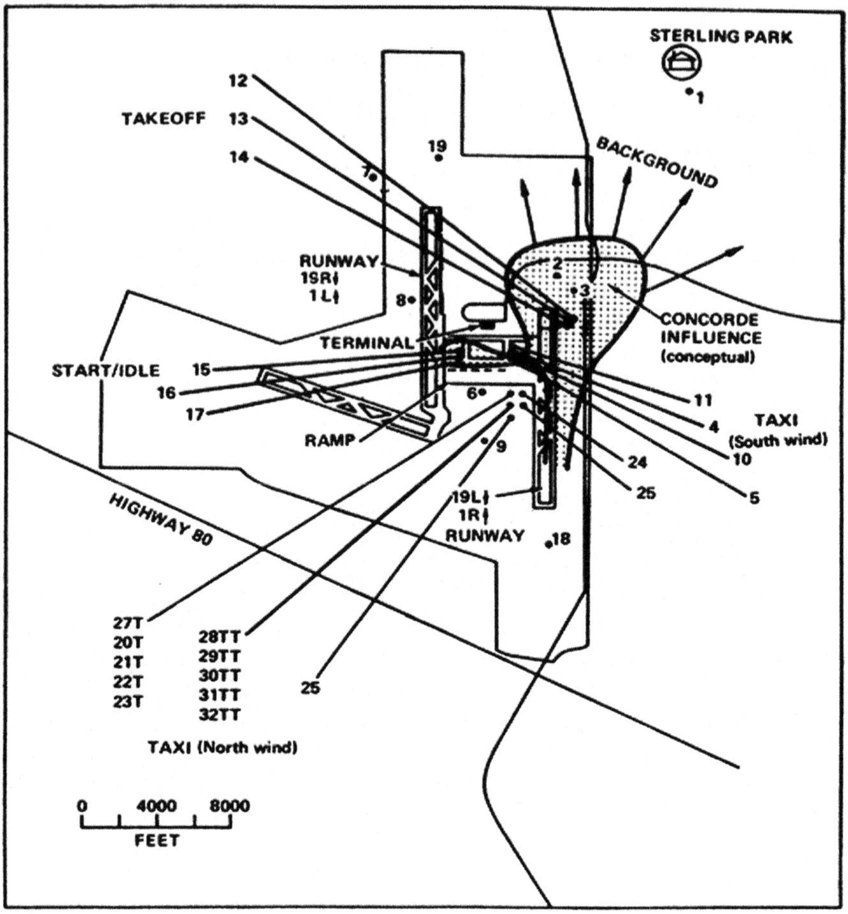


**Fig. 17.1** Air pollution monitoring sites at IAD [1]

## 17.2 Issue Status

The *area of influence* or a *hot spot* problem (worst-case measurement data) is similar to the *area of exceedance* approach from the current local air quality (LAQ) metric task of the CAEP/MDG studies. Measured and calculated concentrations represent 1 h average concentrations. As shown in Fig. 17.1, the impact of a single Concorde is quite low. The concentration envelope (*area of influence*) is bounded by the 1.0 ppm isopleth that represents a small fraction of the 1 h standard of 35 ppm for the CO. The impact of Concorde emissions on the airport from a single aircraft was detected at the end of runways, which are usually more than 1500 feet from the terminal. This large distance between the source and receptor permits aircraft emissions to dilute

**Table 17.1** Emission rate for one engine due to Emission Data Bank

| (a) Olympus-593-610 | | | | (b) JT3D-3B | | | |
|---|---|---|---|---|---|---|---|
| Mode | ERCO, g/s | ERHC, g/s | ERNOx, g/s | Mode | ERCO, g/s | ERHC, g/s | ERNOx, g/s |
| T/F | 184.59 | 18.45 | 60.46 | T/F | 1.76 | 4.696 | 14.2054 |
| Climb | 46.347 | 3.958 | 21.66 | Climb | 2.61 | 1.864 | 9.2268 |
| Approach | 61.95 | 13.35 | 4.09 | Approach | 8.48 | 1.384 | 1.6608 |
| Idle | 42.14 | 14.06 | 0.71 | Idle | 13.23 | 15.12 | 3.375 |

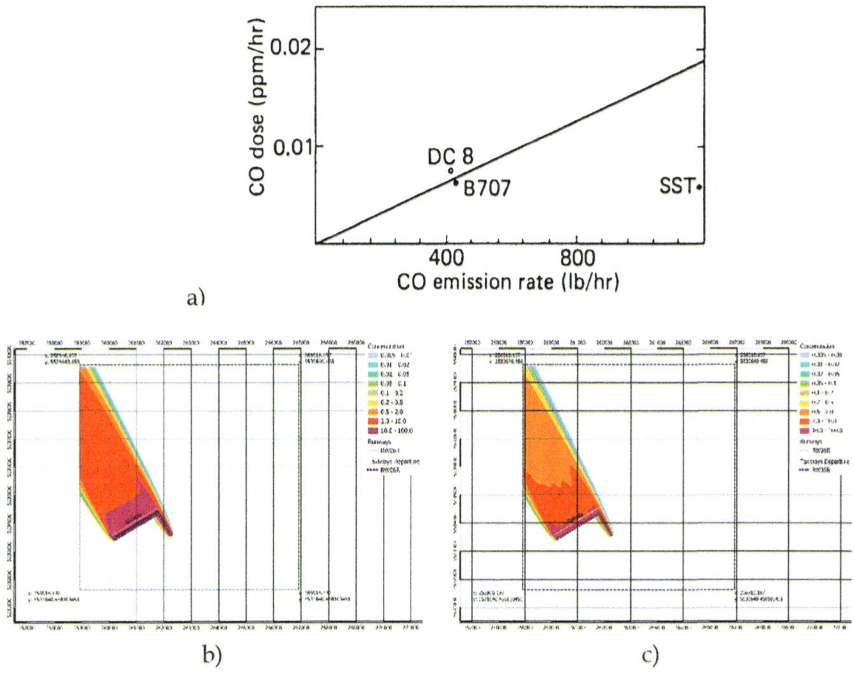


**Fig. 17.2** Averaged per hour CO concentrations via emission rate: (**a**) data from the EIA Concorde program [1]; (**b**) calculated air pollution for Boeing-707; (**c**) calculated air pollution for Concorde

appreciably before reaching the terminal, thereby substantially reducing passenger exposure to aircraft pollution.

When the measured air quality concentrations of Concorde-related CO were compared to the similar other comparable aircraft for which a large number of measurements have been made (Boeing-707 and DC-8), the results imply that the use of emission rate data alone to calculate air quality impacts may be suspected. While the published emission rate for Concorde is approximately three times higher than the emission rates for Boeing-707 and DC-8 (Table 17.1), the measured concentrations of Concorde and these two aircraft were about the same (Fig. 17.2).

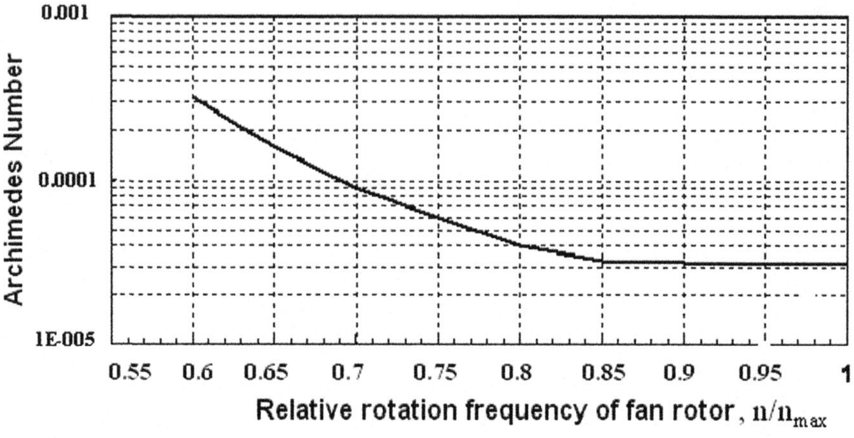


**Fig. 17.3** Archimedes factor as a function of engine power setting (in dependence on fan rotor rotation frequency)

This apparent overestimation of the effects of Concorde emissions could be caused by inaccurate engine test data for this afterburning engine aircraft, different engine–airplane geometry, different engine idle power settings, or other factors. Among other factors, the most important is engine jet height due to buoyancy rise over the ground surface. But irrespective of the cause of this inconsistency, Concorde appears to be affecting ambient air approximately three times less than one would estimate from published source emission rates.

Buoyancy rise is caused by Archimedes forces defined by excess of temperature of jet exhaust gases $T_0$ over the air atmosphere temperature $T_A$. For an estimation of the buoyancy characteristics, the Archimedes number $Ar_0 = gd_0(Q_T - 1)/U_0$ is used, which can be performed as a function of aircraft engine power setting (see Fig. 17.3); $d_0$, exhaust nozzle diameter; $g$, gravitational constant; $Q_T = T_0/T_A$. The data for Archimedes number for Boeing-707 are similar to the data shown in Fig. 17.3 (defined for the similar engines installed on Tupolev-154, Tupolev-134, and Iljushin-62M). But for the Olympus-563 engine of Concorde, these values are much higher, thus the height of their exhaust jets is much higher too, providing similar concentrations as measured during the Concorde EIA campaign at the IAD airport.

## 17.3 Research Results

A model PolEmiCa was developed from measurement data using simple calculations of initial source concentrations and dilution of these concentrations with downwind distance for each mode of aircraft operation [2]. During the IAD campaign, "E" stability and a 1 m/s wind speed were defined and further used in calculating air quality by PolEmiCa because they best fit the worst-case measurement data (a *hot spot* problem). The factors contributing to this difference between the engine emission rates and measured air quality concentration include engine–airplane geometry, engine exhaust temperature (or Archimedes number as in Fig. 17.3), and wake dynamics. Concorde is considerably different from other aircraft (subsonic) in this respect. This issue is still important for the new supersonics.

The calculation test description is presented for the simplified airport case (flight segment scenarios are shown in Fig. 17.2b, c). For the results of the analysis of the studies made by CAEP/MDG for the airport LAQ metric and the Concorde EIA in US airports, three simplified cases are defined:

- For the assessment of queuing and low-speed ground run prior to takeoff of the aircraft in stationary (moving speed equal to zero) conditions with two aside engine operation modes—idle and maximum: Emission is considered for CO (nvPM) and NOx accordingly; engine jet performances are included due to their velocities and temperatures at engine exit; engine jet installations are defined by aircraft designs in consideration; wind direction is the same as the direction of the engine jet; surface covering is a small grass; and Pasquill atmosphere stability class D (E).
- A moving source case is considered for the assessment of taxing aircraft with the engine in idle mode: Emission for CO (nvPM) and engine jet performances should be defined for idle operation mode; wind direction is at an angle of 90° to the direction of the engine jet; surface covering is a small grass; and Pasquill atmosphere stability class D (E).
- A moving source case is considered for the assessment of taking off aircraft along the runway with the engine in maximum mode: Emission for NOx and engine jet performances should be defined for maximum operation mode; wind direction is at an angle of 90° to the direction of the engine jet and aircraft run; wing vortices should be included in consideration for preliminary dilution of engine emission; surface covering is a small grass; and Pasquill atmosphere stability class D (E).

## 17.4 Conclusions

1. Very new engine—different from the previous Olympus of Concorde—will be installed, evidently with smaller temperatures and speeds of the exit jets, so the buoyancy effect of the jets is expected to be much less and closer to the engines of subsonic aircraft. Hence, the emission indexes are expected to be much less than measured for the Olympus engine. These details provide the expectation that the LAQ impact assessment for the new supersonic airplane should be less than that in case of Concorde.
2. During Concorde EIA studies, the emphasis was on CO and NOx measurements so the *influence area* was concentrated around the taxing area in the airport.

Today, the nvPM are the subject of huge concern—for it, the “influence area” will transfer close to the runway, where the emission rate of the nvPM is higher. Hence, the dominance of the NOx emission and appropriate air pollution from the aircraft during their takeoff or/and landing.

3. Three simplified calculation scenarios are considered enough for LAQ assessment of the new aircraft compared with their existing analogs.

**Acknowledgments** We acknowledge EC SENECA project [(LTO) noi*S*e and *E*missio*N*s of sup*E*rsoni*C* Aircraft] no. H2020-MG-2020-SingleStage-INEA-101006742.

## References

1. Howard Segal (1978) Concorde Emissions at Dulles International Airport: Results of Measurements, Journal of the Air Pollution Control Association, 28:10, 1039–1042, [https://doi.org/10.1080/00022470.1978.10470704].
2. Zaporozhets O., Fröhlich J., Stiller J., Synylo K. Improvement of Airport Local Air Quality Modeling. Journal of Aircraft, AIAA. 2017. Vol. 54. – № 5. – P. 1750 –1759 (SCOPUS).

# Chapter 18
# Aircraft Emission and Fuel Burn Assessment Scenarios at Local and Global Levels

**K. Synylo, O. Zaporozhets, A. Krupko, K. Kazhan, V. Makarenko, V. Tokarev, and S. Karpenko**

## 18.1 Introduction

Aviation always brings economic and connectivity benefits (measured in passenger kilometers flown). Forecast results have highlighted the air traffic demand without a proportionate increase in environmental impact that is caused by the improvement in fuel efficiency driven by the introduction of new aircraft, removal of older aircraft, and improvement in operation practice (Fig. 18.1) [1].

However, diverging trends of passenger kilometers flown and aviation emission (due to IMPACT simulations) are expected to grow further in the next 20 years and future technology improvements and unlike to restrain the future traffic growth. Under the base air traffic forecast and assuming an advanced NOx technology improvement rate, emissions would reach around 920,000 tones in 2035 (+42% compared to 2005) (Fig. 18.2) [1].

Current and future technological developments to achieve ACARE 2050 goals involve both the design of the aircraft (sweptback wing) and the design of the engine (high gas speed from a jet engine) to provide high cruise speed, small fuel consumption, and emission reduction. Long-term programs are concentrating on the improvement of engine efficiency via the application of new combustion chamber technologies, and sensor and cooling technologies that permit the engines and auxiliary power units to save 15% of fuel [1].

The development of the scenario for the trends assessment of emission and fuel consumption is based on the combination of flight operations (reference year and

K. Synylo (✉) · A. Krupko · K. Kazhan · V. Makarenko · V. Tokarev · S. Karpenko
National Aviation University, Kyiv, Ukraine
e-mail: vitmakarenko@nau.edu.ua

O. Zaporozhets
Institute of Aviation, Warsaw, Poland

 
T. H. Karakoc et al. (eds.), *The Future of Electric Aviation and Artificial Intelligence*,
Sustainable Aviation, https://doi.org/10.1007/978-3-031-62094-2_18

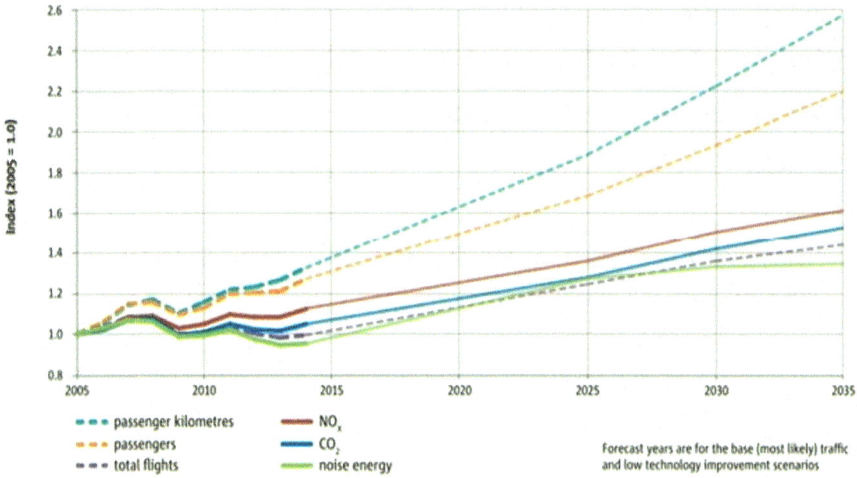


**Fig. 18.1** Noise and emissions forecast to grow slower than passenger kilometers

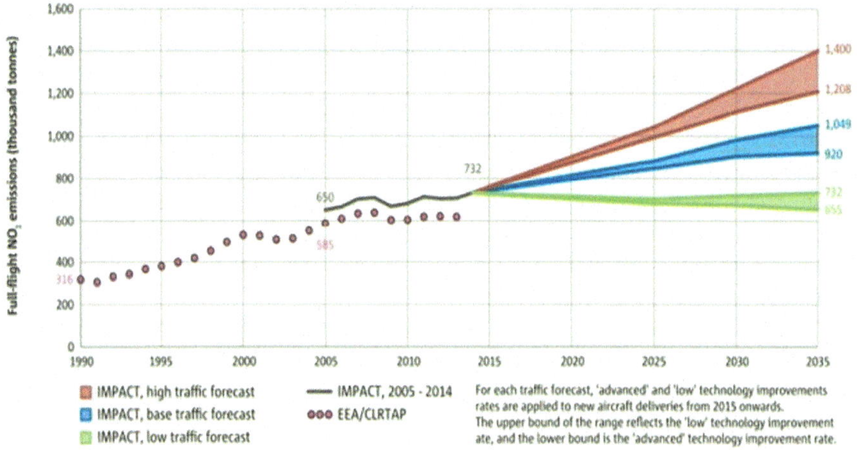


**Fig. 18.2** $NO_X$ emissions are likely to increase in the future, but advanced engine combustor technology

forecast demand) and fleet (active and forecasting). Operations input is grouped into district aircraft types and representative distances and used as a proxy to cover the entire set of base year and future operations. Future scenario includes fleet evolution modeling, particularly novel aircraft needed to replace retiring aircraft (Fig. 18.3) [2].

Environmental modeling in part of greenhouse gas and local air quality is based on the tools recommended by ICAO Doc 9889 [3].

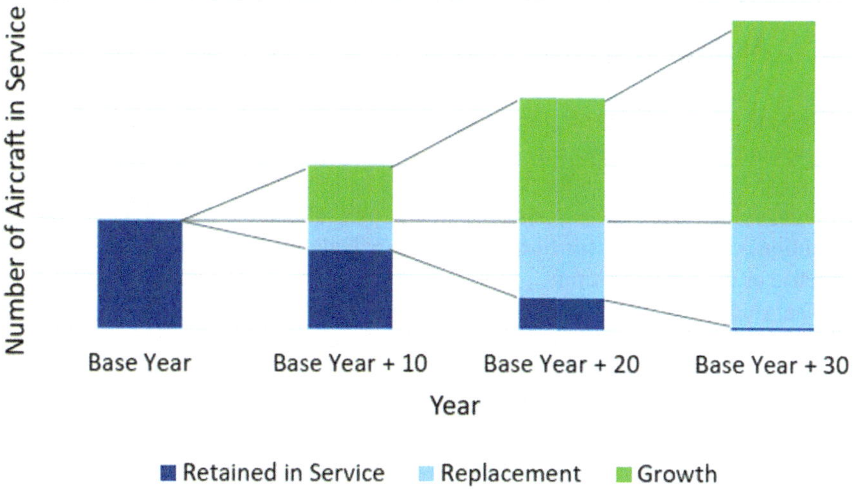


**Fig. 18.3** Fleet evolution modeling

## 18.2 PolEmiCa: Tool for Local and Global Aircraft Emission Inventory

The main purpose of PolEmiCa is to provide the dispersion (pollution) and inventory (emission) calculations for the aircraft engine emission at local and global levels. It includes the aircraft emission from start-up procedures, auxiliary power unit (APU), and ground support equipment (GSE) [4, 5]. The MDG group CAEP successfully completed the verification procedure of the PolEmiCa model in 2019.

Both BFFM2 and P3T3 methods were implemented in the PolEmiCa model to assess emission indexes under real operational (thermodynamic parameters—P3, T3, and fuel flow rate) and meteorological conditions during cruise flight and landing–takeoff cycle. For the estimation of emission rate of $CO_2$ and $H_2O$, PolEmiCa used the fuel consumption method; assuming a mean fuel composition of 13.84% hydrogen by mass, this results in an emission index of 3.15 for $CO_2$ and 1.25 for $H_2O$ (which is the value implemented in the EU Emission Trade System for aviation). Subsequently, each 1 unit of fuel requires 3.4 units of oxygen for complete combustion.

In the past 2 years, PolEmiCa was improved in part regarding emissions inventory (local/global scales), meteorological parameters estimation, and local air quality tasks. Currently, CAEP/12-MDG-FESG/4/WP/09 [5] initiated a feasibility study to test how local air quality (LAQ) dispersion calculations for individual airports can be set up and carried out within MDG in order to provide a quantity that is based on near-ground concentration at and around airports caused by aircraft traffic. A complex model PolEmiCa [6] was involved in the LAQ feasibility study toward a concentration-based LAQ metric.

## 18.3 Global Scenario for Emission and Fuel-Burned Assessment

Global air quality (GAQ) is focused on emission assessment for air traffic routes (cruise emission inventory analysis and its effect on global issues like climate change and cruise flight level). Input data of flight profile with segments (Mach number, altitude, velocity, fuel flow rate, atmospheric pressure, and air temperature for ISA conditions) are provided for further emission calculation for reference and renewed (electric or hydrogen) aircraft.

The study of global emission assessment was done for conventional AN26 (turboprop) and hybrid AN26. The flight range was 2500 km (Fig. 18.4). Operational and meteorological input data for each segment of flight profile were collected by PolEmiCa to calculate average EINOx for real fuel flow at each segment.

Figures 18.5 and 18.6 demonstrate that the fuel flow rate and averaged EINOx for hybrid AN26 are quite less, respectively, 40% and 30% compared with conventional AN26. The same trends were found in the assessed values of fuel burned (kg) and emission NOx (kg) for hybrid AN26 and conventional AN26.

## 18.4 Scenario for Local Air Quality Assessment

For the results of the analysis of the studies made by CAEP/MDG for the airport LAQ metric and Concorde EIA in US airports, three simplified cases are defined:

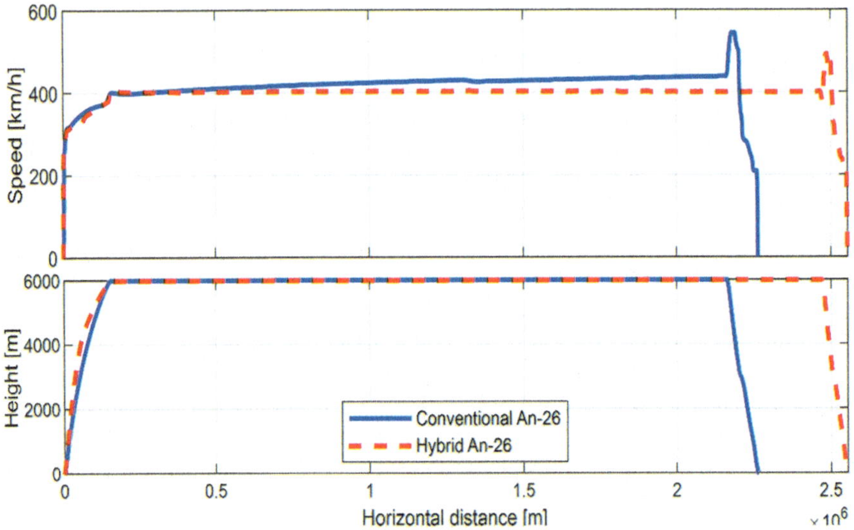


**Fig. 18.4** Flight profile for the conventional AN26 and hybrid AN26

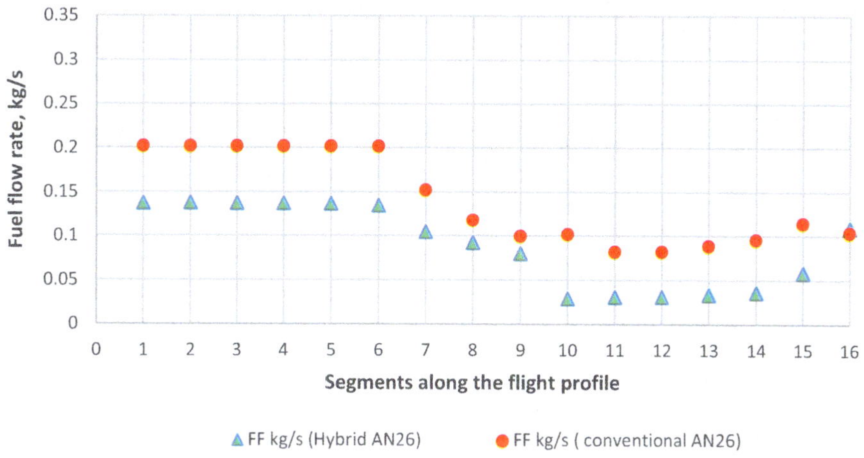


**Fig. 18.5** Estimation of fuel consumption for AN26 (conventional) and hybrid aircraft along the flight profile

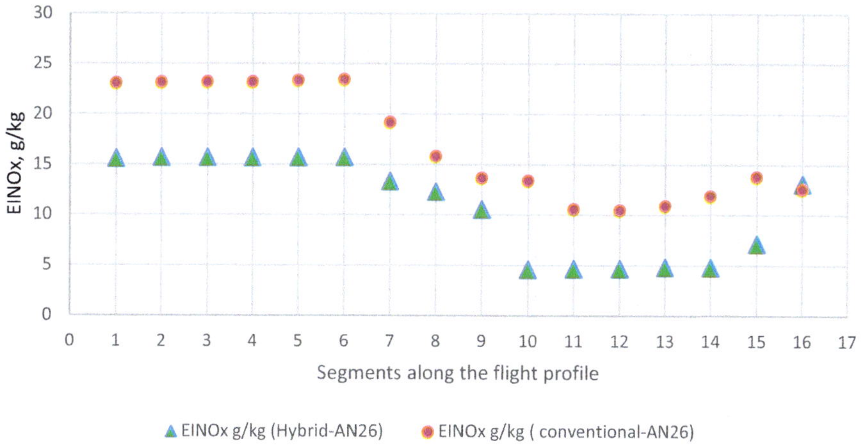


**Fig. 18.6** Estimation of EINOx for AN26 (conventional) and hybrid aircraft along the flight profile

- For the assessment of *queuing and low-speed ground* run prior to takeoff of the aircraft in stationary (moving speed equal to zero) conditions with two aside engine operation modes—*idle and maximum*: Emission is considered for CO (nvPM) and NOx accordingly. Engine jet performances are included due to their velocities and temperatures at engine exit; engine jet installations are defined by aircraft designs in consideration; wind direction is the same as the direction of the engine jet; surface covering is a small grass; and Pasquill atmosphere stability class D (E) for all cases.
- A *moving source* case is considered for the assessment of taxing aircraft with the engine in idle mode: Emission for CO (nvPM) and engine jet performances

should be defined for idle operation mode; wind direction is at an angle of 90° to the direction of the engine jet; and surface covering is a small grass.

- A *moving source case* is considered for the assessment of taking off aircraft along the runway with the engine in maximum mode: Emission for NOx and engine jet performances should be defined for maximum operation mode. Wind direction is at an angle of 90° to the direction of the engine jet and aircraft run; wing vortices should be included in consideration for preliminary dilution of engine emission; and surface covering is a small grass.

Local air quality assessment should take into account infrastructural needs in support of updated fleet.

## References

1. Perspectives for aeronautical research in Europe /Chapter 18. Decarbonisation of Aviation by 2050, Final report of EU Project within Horizon 2020, 2021 1
2. Consolidated statement of continuing ICAO policies and practices related to environmental protection A39-1 – General provisions, noise and local air quality; A39-2 – Climate change; A39-3 – Global Market-based Measure (MBM) Scheme, ICAO, Montreal, 2016. 2
3. ICAO Doc 9889. Airport Air Quality. – 1st ed. -2011. – 200p.
4. Zaporozhets O., Fröhlich J., Stiller J., Synylo K. Improvement of Airport Local Air Quality Modeling // Journal of Aircraft, AIAA. -2017. - Vol. 54. – № 5. – P. 1750 –1759 (SCOPUS).4
5. CAEP/12-MDG-FESG/4-WP/09: Feasibility study for a LAQ metric – work progress // Fourth Meeting – Virtual 15 to 19 June 2020 5
6. Zaporozhets O. PolEmiCA Local Air Quality Model Evaluation: Inventory/Dispersion Assessment & Analysis / O. Zaporozhets, K. Synylo // CAEPMDG9 Meeting, Cologne, 1–2 May 2015. – 41 6

# Chapter 19
# Noise Assessment Scenarios for New Airplane Climate Efficient Designs

**Kateryna Kazhan, Vitalii Makarenko, Vadym Tokarev, Kateryna Synylo, Oleksander Zaporozhets, Andrzej Chyla, and Michal Bukala**

## 19.1 Introduction

Over the last decades, the increasing globalization and the associated need for substantially shortened travel times have led to public and privately funded development of supersonic aircraft, mainly in the USA. In view of the soon-to-be-expected market entry of the supersonic business jet Aerion AS2 [1] and existing plans for commercial SST aircrafts with a payload of more than 100 passengers, the International Civil Aviation Organization (ICAO) needs to define standards for the advent of this new generation of supersonic aviation.

## 19.2 Main Achieved Results from Concorde Studies

This Concorde Environmental Impact Assessment (EIA) study at Dulles International Airport demonstrated that the contribution of aircraft emissions to air quality occurs mainly during only two operational modes: *queuing* and *low-speed ground run prior to takeoff* [2].

Single-event measurements were used to define the Concorde influence area referred to in Fig. 19.1 and provide the detailed data for background pollution analysis. The measurement procedure was to place three monitoring stations

K. Kazhan (✉) · V. Makarenko · V. Tokarev · K. Synylo
National Aviation University, Kyiv, Ukraine
e-mail: vitmakarenko@nau.edu.ua

O. Zaporozhets
Warsaw Institute of Aviation, Warsaw, Poland

A. Chyla · M. Bukala
NOISE ACH, Warsaw, Poland

 
T. H. Karakoc et al. (eds.), *The Future of Electric Aviation and Artificial Intelligence*, Sustainable Aviation, https://doi.org/10.1007/978-3-031-62094-2_19

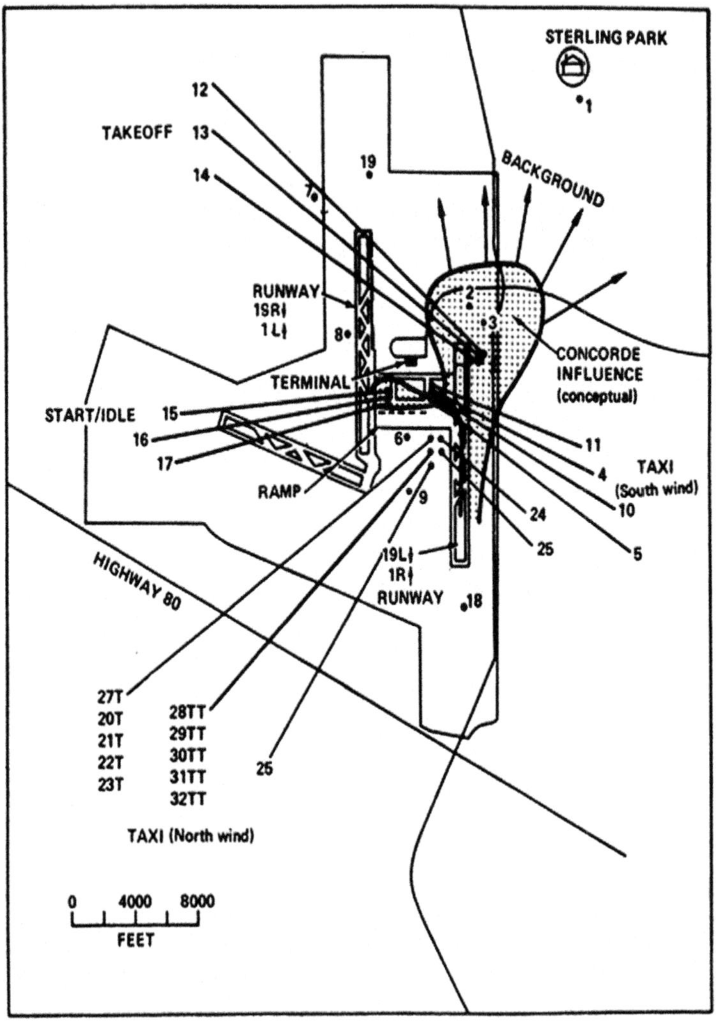


**Fig. 19.1** Air monitoring sites at Dulles International Airport

downwind of the Concorde movement path (*started, taxied, and took off*) and measure the dispersion of the emission plume as it was transported by the wind over these stations. The concept of the "area of Concorde influence" introduced the emissions blend to the lowest level normally detectable above background.

A model was developed from measurement data using simple calculations of initial source concentrations and dilution of these concentrations with downwind distance for each mode of aircraft operation. "E" stability and a 1 m/s wind speed were used to calculate air quality because they best fit the worst-case measurement data (a "hot spot" problem). Winds from the south were assumed to ensure the shortest transport distance between the jet ramp and the terminal and also to ensure

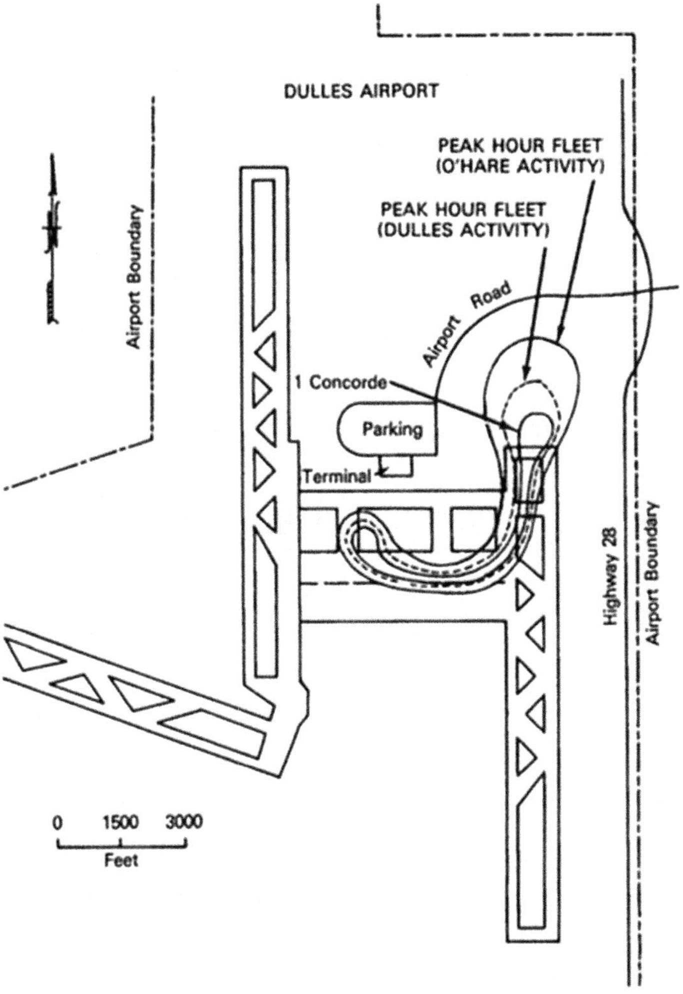


**Fig. 19.2** CO area of influence: isopleths represent 1 ppm concentration

maximum pollution build-up along the north/south taxiway. Pollution values given in Fig. 19.2 represent calculated 1 h average concentrations. The concentration envelope ("area of influence") is bounded by the 1.0 ppm isopleth, which represents a small fraction of the 1 h standard of 35 ppm. As shown in Fig. 19.2, the impact of one Concorde aircraft is quite low.

The impact of Concorde emissions on the airport from a single aircraft is detected at the end of runways that are usually more than 1500 feet from the terminal. This large distance between the source and the receptor permits aircraft emissions to dilute appreciably before reaching the terminal, thereby substantially reducing passenger exposure to aircraft pollution.

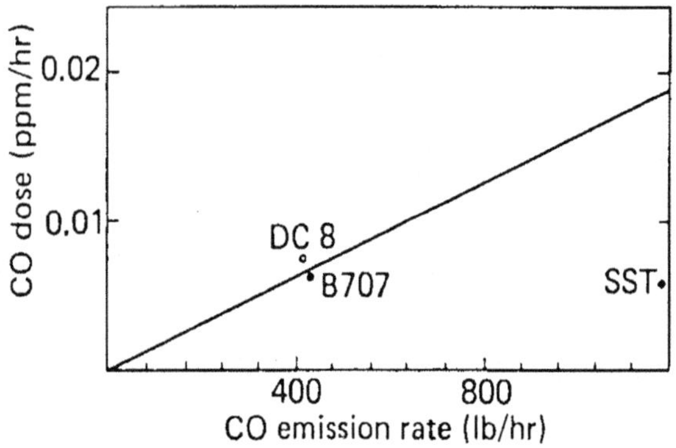


**Fig. 19.3** CO dose via emission rate

The "*area of influence*" or a "*hot spot*" problem (worst-case measurement data) are similar to the "*area of exceedance*" approach from the current Local Air Quality (LAQ) metric task of the Committee on Aviation Environmental Protection/ Modeling and Databases Group (CAEP/MDG) studies.

When the measured air quality effects (doses) of Concorde-related CO are compared to the doses of other comparable aircraft for which a large number of measurements have been made (707 and DC-8), the results imply that the use of emission rate data alone to calculate air quality impacts may be suspected. While the published emission rate for Concorde is approximately three times higher than the emission rates for 707 and DC-8, the measured doses of Concorde and 707 and DC-8 are about the same. This apparent overestimation of the effect of Concorde emissions could be caused by inaccurate engine test data for this afterburning-engine aircraft, different engine–airplane geometry, different engine idle power settings, or other factors. But irrespective of the cause of this inconsistency, the Concorde appears to be affecting ambient air approximately three times less than one would estimate from published source emission rates (Fig. 19.3).

## 19.3 Recommendations and Conclusions

Engine emission rate measurements may not reflect aircraft environmental impact. Factors contributing to this difference between published engine emission rates and measured air quality concentration include engine–airplane geometry, engine exhaust temperature, and wake dynamics. Concorde is considerably different from other aircraft in this respect.

The most direct measure of aircraft emissions impact is the actual ambient air quality change as measured near the engine emission's source; thus, the ambient air measurements obtained in the current program are an important supplement to engine emission measurements made in the environment of the engine test stand. In this section, the calculation test description is presented for the simplified airport case (flight segment scenarios). For the results of the analysis of the studies made for Concorde EIA in US airports, three simplified cases are defined:

- For the assessment of *queuing* and *low-speed ground run prior to takeoff* of the aircraft in stationary (moving speed equal to zero) conditions with two aside engine operation modes—*idle* and *maximum*: Emission is considered for CO "non-volatile" particular matter (nvPM) and NOx accordingly; engine jet performances are included due to their velocities and temperatures at engine exit; engine jet installations are defined by aircraft designs in consideration; wind direction is the same as the direction of the engine jet; surface covering is a small grass; and Pasquill atmosphere stability class D (E).
- A moving source case is considered for the assessment of taxing aircraft with the engine in idle mode: Emission for CO (nvPM) and engine jet performances should be defined for idle operation mode; wind direction is at an angle of 90° to the direction of the engine jet; surface covering is a small grass; and Pasquill atmosphere stability class D (E).
- A moving source case is considered for the assessment of taking off aircraft along the runway with the engine in maximum mode: Emission for NOx and engine jet performances should be defined for maximum operation mode; wind direction is at an angle of 90° to the direction of the engine jet and aircraft run; wing vortices should be included in consideration for preliminary dilution of engine emission; surface covering is a small grass; and Pasquill atmosphere stability class D (E).

## References

1. Final report: Perspectives for Aeronautical Research in Europe, Chapter.4, pg. 146: https://www.pareproject.eu/publications
2. Howard Segal (1978) Concorde Emissions at Dulles International Airport: Results of Measurements, Journal of the Air Pollution Control Association, 28:10, 1039–1042, [https://doi.org/10.1080/00022470.1978.10470704]

# Chapter 20
# Composite Gradient Coatings Design for Ensuring Electromagnetic Compatibility of Onboard and Ground Electronic Equipment

**V. Glyva, L. Levchenko, N. Burdeina, Y. Biruk, and O. Tykhenko**

## 20.1 Introduction

With the increase of the electrical and electronic equipment onboard aircraft and the dense placement of those equipments at the small areas increase of the probability of inerference and incorrect functioning is higher than the designated use of the frequency spectrum due to the mutual electromagnetic band use. This requires the implementation of equipment protection measures. The most effective means is the shielding of the fields of individual sources and the protection of certain zones and premises from electromagnetic influences by using electromagnetic shields of the required configuration.

## 20.2 Issue Status

The fundamental disadvantage of all metals is the large coefficient of electromagnetic wave reflection. This worsens the electromagnetic environment in unprotected areas. Metal screens are heavy and inconvenient for complex surface configurations.

V. Glyva (✉) · N. Burdeina · Y. Biruk
Kyiv National University of Construction and Architecture, Department of Physics, Kyiv, Ukraine
e-mail: burdeina.nb@knuba.edu.ua

L. Levchenko
Igor Sikorsky Kyiv Polytechnic Institute, Department of Digital Technologies in Energy, Kyiv, Ukraine

O. Tykhenko
National Aviation University, Department of Ecology, Kyiv, Ukraine



T. H. Karakoc et al. (eds.), *The Future of Electric Aviation and Artificial Intelligence*, Sustainable Aviation, https://doi.org/10.1007/978-3-031-62094-2_20

The greatest attention is paid to the development of composite protective materials. Research by Kolcunová et al. (2020) and Ji et al. (2021) shows the development of materials and the results that shield electromagnetic fields due to a special surface structure. Such designs are effective for flat surfaces and rooms, but are inconvenient for complex shapes and applications onboard aircraft. The results of shielding materials and the developments based on polymer and nanoparticle use are presented by Mondal et al. (2017); Yadav et al. (2019). But the nanomaterials used have a high cost, and a thin single-layer coating is not effective enough. Studies by Butenko et al. (2019) and Senyk et al. (2020) are related to the development of liquid-based compositions. The advantages are the possibility of application on the surface of any shape and high efficiency in the high-frequency region of the electromagnetic spectrum. The efficiency is provided by the high conductivity of the filler made of graphitized carbon black and graphite linear structures. But such materials are ineffective for stationary and quasi-stationary fields. In addition, they are multicomponent, which increases the risk of material degradation during long-term use.

Based on the fundamental relations of continuous media of the electrodynamics, it can be expected that the approach impedance of the surface layer of the composite to the indicator of propagation electromagnetic waves in the medium will reduce their reflection coefficient. But in this case material shielding coefficient also decreases. Its increase is possibly due to the creation of a multilayer structure with layers having different electrophysical and magnetic properties. It is advisable to implement it on the basis of liquid materials (matrices) with a small thickness of each layer. Such structures are convenient for application on a complex configuration of the surface, have low weight, and can provide shielding of electromagnetic fields of a wide frequency range and stationary (quasi-stationary) fields. On the basis of such structures, it is possible to develop coatings that protect electronic equipment from external electromagnetic influences and reduce field levels inside moving objects due to the absorption of fields' electromagnetic energy generated by onboard electrical and electronic equipment.

## 20.3 Research Results

The main problem in the design of composite electromagnetic shields of the required efficiency is the lack of stationary reference data on the electrophysical and magnetic properties of most composites. Therefore, to design electromagnetic protection, at the first stage, it is necessary to determine the amplitude–frequency characteristics of the field that needs to be shielded. Based on these data, the constructive solution and the composition of materials are selected. Liquid-based mixtures (varnish, liquid polymer) have a small thickness of 50–200 μm. Therefore, their protective properties, especially for the field of a wide frequency band, are unsatisfactory. It is advisable to create a multilayer structure with different properties of each layer. The outer layer should have a wave resistance comparable to air resistance (377 Ω), which ensures minimum reflection coefficients.

Surface resistance is determined by the ratio $\sqrt{\mu/\varepsilon}$, where $\mu$ and $\varepsilon$ are magnetic and dielectric relative permeability. Liquid materials suitable for use as a matrix (varnishes, paints, latex) have a dielectric constant from 4 (XC varnish) to 24 (latex), a magnetic permeability close to 1. Therefore, a magnetic material is added to them in dispersed form (e.g., magnetite). It is advisable to determine the necessary content of impurities using the Lorentz ratio

$$\mu = 1 + \frac{\upsilon_{\mathrm{m}}(\mu_{\mathrm{m}} - 1)}{1 + \frac{1 - \upsilon_{\mathrm{m}}}{3}(\mu_{\mathrm{m}} - 1)},$$

where $\mu$ is the effective magnetic permeability of the magneto-dielectric, $\mu_{\mathrm{m}}$ is the magnetic permeability of the filler, and $\upsilon_{\mathrm{m}}$ is the volume content of the filler in the material.

This ratio takes into account the volumetric content of impurities. It is convenient to use weight units for the preparation of the mixture. Therefore, the conversion of weight ratios into volume is carried out according to the ratio

$$\frac{V_{\mathrm{m}}}{V_{\mathrm{d}}} = \frac{k\rho_{\mathrm{m}}}{\rho_{\mathrm{m}}},$$

where $V_{\mathrm{m}}$, $V_{\mathrm{d}}$ are the volumes of filler and dielectric; $\rho_{\mathrm{m}}$, $\rho_{\mathrm{d}}$ are the density of filler and dielectric; and $k = \frac{P_{\mathrm{m}}}{P_{\mathrm{d}}}$ is the weight ratio of filler and dielectric.

To determine the dielectric constant of the composite, it is advisable to use the Odelevsky formula:

$$\varepsilon = \varepsilon_{\mathrm{d}}\left[1 + \frac{\upsilon_{\mathrm{m}}(\varepsilon_{\mathrm{m}} - \varepsilon_{\mathrm{d}})}{\left(1 - \frac{\upsilon_{\mathrm{m}}}{\upsilon_{\mathrm{k}}}\right) \times F \times (\varepsilon_{\mathrm{m}} - \varepsilon_{\mathrm{d}}) + \varepsilon_{\mathrm{d}}}\right],$$

where $\varepsilon_{\mathrm{d}}$, $\varepsilon_{\mathrm{m}}$ are the dielectric constants of the matrix and the filler, $\upsilon_{\mathrm{m}}$ is the volume content of the filler, $\upsilon_{\mathrm{k}}$ is the critical volume content of the filler at which the shielding particles come into contact with each other, and $F$ is the depolarization coefficient:

$$F = \ln\left(\frac{2l}{d}\right) / \left(\frac{l}{d}\right)^2,$$

where $l$ is the average particle length and $d$ is the average diameter.

$$\upsilon_k = \frac{[3F(1-F)]^{0,6}}{\left[4F\sqrt{l/d} + (1+F)/\sqrt{l/d}\right]^{0,6}}$$

The predominant ratio of the length and diameter of the impurity particles can be determined by microscopic studies. In many cases, for example, for flaky graphite, this indicator is approximate, so the calculations give an approximate result. Therefore, the smallest calculation error is for filler particles of spherical or similar shape. But in this case the concentration of the filler in the inner layer should be greater. This is explained by the fact that a sharp increase in the absorption coefficient occurs due to the appearance of the percolation effect—an increase in the conductivity of the composite due to the formation of conductive linear structures. Such structures are formed due to the contact of conductive components.

In general, the conductivity of material layers should increase from the outer layer to the inner layer. Experiments on shielding coefficients were carried out for a three-layer structure based on geopolymer paint and magnetite with a content of 30%, 45%, and 60% (by weight). The average size of magnetite particles was 12 μm. The thickness of each layer did not exceed 70 μm. The reflection coefficient of the three-layer structure for an electromagnetic field with a frequency of 2.45 GHz was 0.12–0.13. The total shielding coefficient due to the absorption of electromagnetic energy was 16.5–17.2, which exceeds the total shielding coefficient of each of the three layers (10, 3). At the same time, the shielding coefficient of a quasi-stationary magnetic field with a frequency of 50 Hz was 14.2–14.6.

In many cases, the external electromagnetic field needs to be shielded and the field generated by electronic equipment in a limited space must be absorbed. For this, the second (inner) layer must have maximum protective properties, and the first and third parameters that minimize the reflection of electromagnetic waves. To increase the shielding efficiency, flaky graphite in the amount of 5% (by weight) was added to the outer layer. This increased the shielding factor of the structure by 7.7–8.1.

## 20.4 Conclusions

1. To increase the levels of electromagnetic compatibility of electrical and electronic equipment of aircraft and ground services, the most effective means is shielding of external and internal electromagnetic fields with protective materials. Compositions based on varnishes, paints (matrix), and magnetic and conductive filler are suitable for application on the surface of any configuration.
2. A three-layer structure with a gradient of electrophysical and magnetic properties allows minimizing the reflection coefficients of electromagnetic waves due to the selection of components that ensure the value of the wave resistance of the surface layer close to the air resistance. Electromagnetic energy is absorbed in the following layers.

3. With a magnetite content of 30%, 45%, and 60% (by weight) in the geopolymer paint matrix, a three-layer structure was obtained (the thickness of each layer did not exceed 70 μm). The reflection coefficient from the first layer (30%) is 0.12–0.13. The overall shielding coefficient is 14.2–14.6 (field frequency 2.45 GHz). To ensure double-sided shielding, it is advisable to add flaky graphite to the inner layer. With its content of 5% (by weight), the shielding coefficient increased by 7.7–8.1. The shielding coefficient of the magnetic field with a frequency of 50 Hz by the three-layer structure was 14.2–14.6, which can be considered satisfactory.

## References

Butenko, O. et al. (2019) 'Pure ultrafine magnetite from carbon steel wastes', *Materials Today: Proceedings*, 6, pp. 270–278. Available at: https://doi.org/10.1016/j.matpr.2018.10.104.

Ji, K.F. et al. (2021) 'Design of Ultra-wideband Low RCS Reflecting Screen Based on Phase Gradient Metasurface', *Radioengineering*, 30(2), pp. 314–322. Available at: https://doi.org/10.13164/re.2021.0314.

Kolcunová, I. et al. (2020) 'Shielding Effectiveness of Electromagnetic Field by Specially Developed Shielding Coating', *Acta Physica Polonica A*, 137(5), pp. 711–713. Available at: https://doi.org/10.12693/aphyspola.137.711.

Mondal, S. et al. (2017) 'Low percolation threshold and electromagnetic shielding effectiveness of nano-structured carbon based ethylene methyl acrylate nanocomposites', *Composites Part B: Engineering*, 119, pp. 41–56. Available at: https://doi.org/10.1016/j.compositesb.2017.03.022.

Senyk, I.V. et al. (2020) 'Development and Application of Thin Wide-Band Screening Composite Materials', *Physics and Chemistry of Solid State*, 21(4), pp. 771–778. Available at: https://doi.org/10.15330/pcss.21.4.771-778.

Yadav, R.S. et al. (2019) 'Polypropylene Nanocomposite Filled with Spinel Ferrite NiFe2O4 Nanoparticles and In-Situ Thermally-Reduced Graphene Oxide for Electromagnetic Interference Shielding Application', *Nanomaterials*, 9(4), p. 621. Available at: https://doi.org/10.3390/nano9040621.

# Chapter 21
# Renewable Energy Systems for Airports and Aerodromes: A Comprehensive Patent Review and Technological Analysis

**Omar Alharasees, Utku Kale, Jozsef Rohacs, and Daniel Rohacs**

## 21.1 Introduction

Renewable energy is crucial for tackling climate change and addressing energy resource depletion. The aviation industry, including airports and aerodromes, has been actively adopting renewable energy systems to reduce greenhouse gas emissions and achieve sustainability goals (Rodríguez-Sanz et al. 2020). This chapter examines seven key renewable energy types (solar collectors, solar photovoltaic, wind energy, wave energy, tidal energy, hydro energy, and geothermal energy) and their application in airports and aerodromes, contributing to the understanding of sustainable energy solutions in the aviation sector.

Solar collectors are increasingly integrated into airports for space heating and cooling (Kilic and Dursun 2017) as demonstrated by Barcelona-El Prat Airport's large-scale solar thermal system (Morganti et al. 2019). This technology significantly reduces conventional energy consumption. Solar photovoltaic systems have also been widely adopted in airports worldwide, with Cochin International Airport serving as the first fully solar-powered airport (Sukumaran and Sudhakar 2017), These successful implementations showcase the aviation sector's progress in harnessing solar energy for sustainable operations.

Wind energy is a promising renewable source for airports (Chowdhury et al. 2021), exemplified by wind turbines at Amsterdam Airport Schiphol (Kılkış and Kılkış 2017), meeting a significant portion of the electricity demand. Coastal airports can also benefit from wave energy converters, as demonstrated by the exploration at Bristol Airport (Halpern and Graham 2013). These advancements highlight the

O. Alharasees (✉) · U. Kale · J. Rohacs · D. Rohacs
Department of Aeronautics, and Naval Architecture, Faculty of Transportation Engineering and Vehicle Engineering, Budapest University of Technology and Economics, Budapest, Hungary
e-mail: oalharasees@edu.bme.hu; kale.utku@kjk.bme.hu; rohacs.jozsef@kjk.bme.hu; rohacs.daniel@kjk.bme.hu

 
T. H. Karakoc et al. (eds.), *The Future of Electric Aviation and Artificial Intelligence*, Sustainable Aviation, https://doi.org/10.1007/978-3-031-62094-2_21

potential for airports to harness wind, wave, and tidal energy for sustainable operations. Airports near water bodies have successfully implemented hydro energy systems, as demonstrated by the Vancouver International Airport's hydroelectric power plant (Su and Karney 2014). Similarly, airports located in geologically favorable regions, such as Keflavík International Airport in Iceland, utilize geothermal energy for heating and cooling purposes (Harrison et al. 1990). These examples showcase the effective utilization of hydro and geothermal energy in airport operations..

This chapter analyzes the patents filed between 2010 and 2022 using the WIPO Espacenet Patent search platform (Espacenet patent search 2023), to identify technological advancements and trends in renewable energy for airports and aerodromes. The study aims to provide insights for policymakers, airport authorities, and researchers, facilitating informed decision-making and promoting the adoption of renewable energy solutions in the aviation industry. This study analyzes patents to assess renewable energy systems for airports and aerodromes, focusing on solar, wind, wave, tidal, hydro, and geothermal energy. It aims to identify the trends, innovations, and integration opportunities. Findings inform policymaking, airport operations, and research, facilitating widespread adoption of sustainable energy and reducing environmental impact.

## 21.2 Method

This research utilizes a systematic methodology to examine the application and future potential of seven renewable energy types in airports and aerodromes. The approach involves reviewing scholarly literature, analyzing patents, and assessing future perspectives. By focusing on solar collectors, solar photovoltaic (PV), wind energy, wave energy, tidal energy, hydro energy, and geothermal energy, this study aims to comprehensively understand their characteristics, practical uses, and potential advancements in airport settings.

The literature review explores academic databases and industry reports to gather knowledge on each renewable energy type. Solar collectors at Hong Kong International Airport reduce reliance on traditional energy sources (Zu and Zhang 2023). Solar PV systems are deployed at Indianapolis International Airport and Cochin International Airport for lighting (Sreenath et al. 2020). Wind energy is harnessed through the integration of wind turbines at Amsterdam Airport Schiphol (Zhou 2022). Bristol Airport explores the utilization of wave energy converters for coastal energy generation.

Micro hydropower installations enable the effective harnessing of hydro energy from flowing or falling water. Frankfurt Airport and Vancouver International Airport have integrated small-scale hydropower systems in nearby rivers or waterways. Geothermal energy, sourced from the Earth's crust, is successfully utilized for heating and cooling at Keflavík International Airport in Iceland.

The second stage involves analyzing patents for practical examples and advancements of each renewable energy type in airports and aerodromes, such as solar collector design, solar PV optimization, and innovative wave and tidal energy technologies. The final stage assesses the future prospects of renewable energy types in the aviation industry, considering advancements, scalability, feasibility, and environmental impact. The goal is to determine their potential for reducing carbon emissions, improving energy efficiency, and promoting sustainability in airports and aerodromes. This research examines seven distinct renewable energy types and their applications in airport settings. The study contributes to the existing knowledge on renewable energy integration in the aviation sector, providing valuable insights from scholarly literature, patent analysis, and future perspectives assessment. The findings offer practical guidance for sustainable energy planning in airport design and operation.

## 21.3 Results and Discussion

This section analyzes renewable energy implementation in airports and aerodromes across eight countries (China, the USA, Japan, Korea, Australia, Canada, and Russia). Solar, wind, and wave energies are prioritized for their availability, scalability, and emissions reduction potential. Figure 21.1 provides a global overview of the seven energy types.

The following sections highlight the findings from Fig. 21.2, which presents an individual comparison of each energy type in these countries, demonstrating the varying levels of adoption and success. Figure 21.3 presents a temporal evaluation of the seven energy types.

Figure 21.2 shows that China leads in solar collectors and photovoltaic (PV) installations, while the USA prioritizes wind energy projects. Japan and

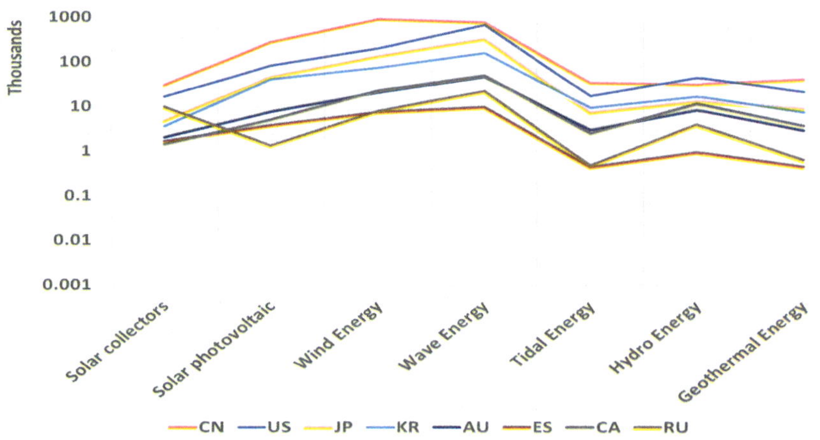


**Fig. 21.1** Comparison of the seven energy types

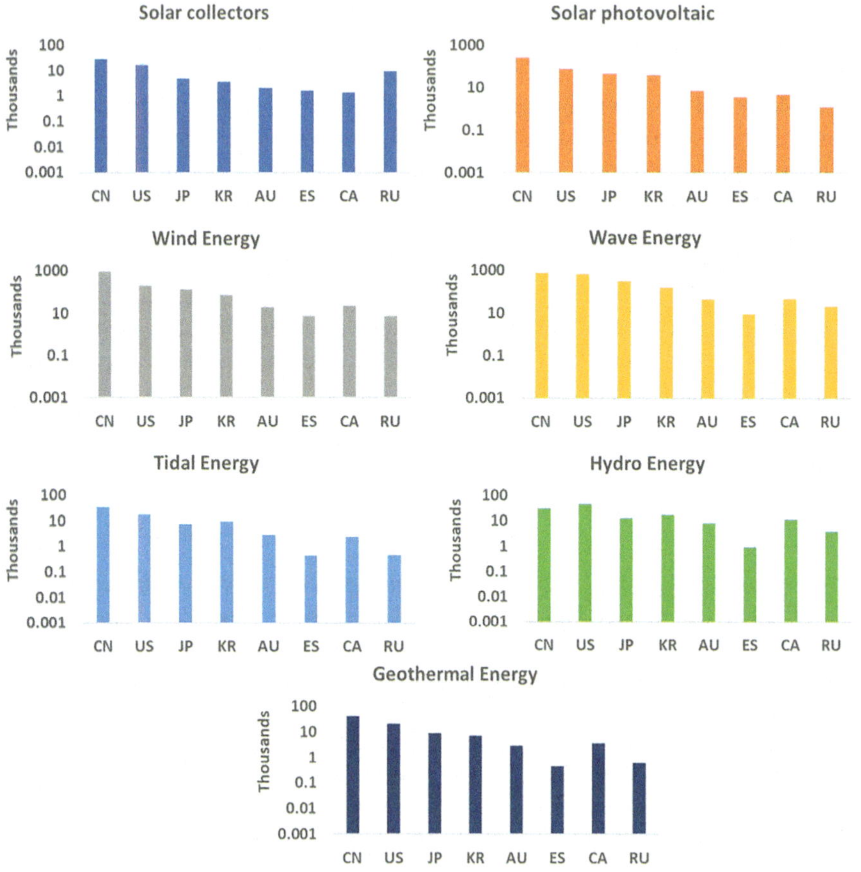


**Fig. 21.2** Individual comparison for each energy type between the eight countries

Korea excel in solar PV technology, Australia focuses on wave energy, and Canada and Russia show interest in hydro and geothermal energy, respectively. Between 2010 and 2022, the mentioned energy types experienced notable growth, reflecting their increasing importance in the aviation sector's sustainability efforts. From Fig. 21.3, we can illustrate the rise in solar PV installations, the expansion of wind energy farms, and the exploration of wave energy technologies, demonstrating the commitment to cleaner energy sources and reduced reliance on fossil fuels. In summary, the analysis of top countries demonstrates a diverse adoption of renewable energy in airports. China excels in solar, while the USA emphasizes wind. Japan, Korea, Australia, Canada, and Russia make notable progress in various renewables. Solar, wind, and wave energies show significant acceleration, highlighting their potential for transforming the aviation industry. Continued investment and research in renewable technologies are crucial for a sustainable future in aviation.

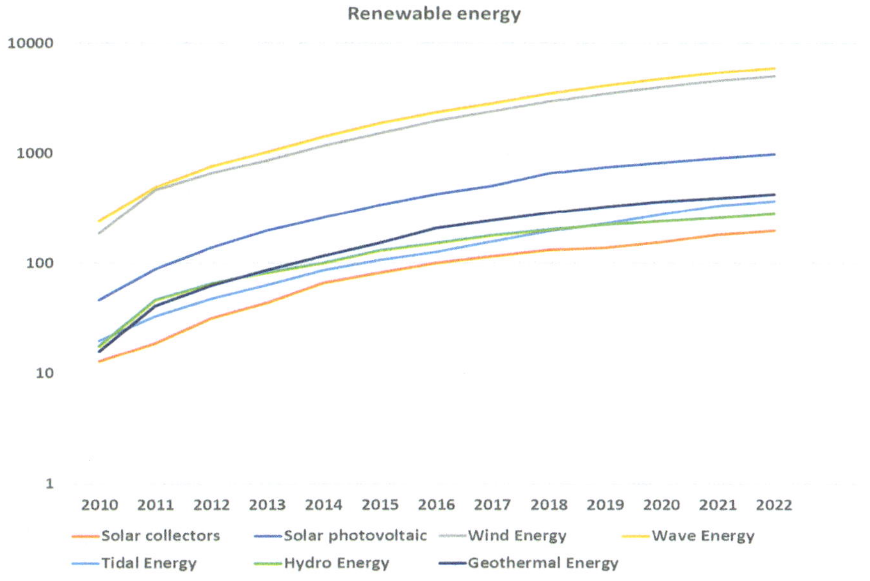


**Fig. 21.3** Renewable energy comparison between 2010 and 2022

## 21.4 Conclusion

This study assesses seven renewable energy types (solar collectors, solar PV, wind energy, wave energy, tidal energy, hydro energy, and geothermal energy) in airports. The analysis reveals their potential for sustainable transformation in the aviation industry. Top countries (China, the USA, Japan, Korea, Australia, Canada, and Russia) exhibit varying levels of adoption and success in implementing these energy types. Solar, wind, and wave energies are prominent and rapidly advancing renewable energy sources in airports. China excels in solar collector and solar PV installations, while the USA leads in wind energy projects. Japan, Korea, and Australia demonstrate notable progress in solar PV and wave energy technologies. Additionally, Canada and Russia show growing interest in hydro and geothermal energy. These trends signify a global commitment to reducing emissions, enhancing energy efficiency, and fostering sustainability in the aviation industry.

The 2010–2022 comparison highlights the rapid growth of solar, wind, and wave energies. Further investment and research are crucial to unlock their full potential in airport operations. Prioritizing these renewables is vital for achieving industry sustainability, reducing fossil fuel dependence, and mitigating environmental impacts. Lessons from leading countries offer valuable guidance for global adoption of renewable energy solutions in aviation infrastructure.

In conclusion, renewable energy in airports and aerodromes shapes a sustainable aviation future. Solar, wind, wave, tidal, hydro, and geothermal energy are key. Collaboration drives global adoption, reducing emissions and enhancing resilience. Implementing renewables in airports is vital for a greener future.

**Acknowledgments** This work was supported by Project No. 2022–2.1.1-NL-2022-00012, which has been implemented with the support provided by the Ministry of Culture and Innovation of Hungary from the National Research, Development and Innovation Fund, financed under the National Laboratories funding scheme.

## References

Chowdhury, H. *et al.* (2021) 'Energy, emission, profitability, and sustainability analyses of a grid-connected solar power plant proposed in airport sites of Bangladesh: a case study', *Environmental Science and Pollution Research*, 28(43), pp. 61369–61379. Available at: https://doi.org/10.1007/S11356-021-14973-5/TABLES/4.

*Espacenet – patent search*. Available at: https://worldwide.espacenet.com/patent/ (Accessed: 13 June 2023).

Halpern, N. and Graham, A. (2013) 'Airport marketing', *Airport Marketing*, pp. 1–225. Available at: https://doi.org/10.4324/9780203117903.

Harrison, R., Mortimer, ND and Smarason, OB (1990) 'GEOTHERMAL DISTRICT HEATING IN ICELAND', *Geothermal Heating*, pp. 468–541. Available at: https://doi.org/10.1016/B978-0-08-040503-2.50023-8.

Kilic, B. and Dursun, E. (2017) 'Integration of innovative photovoltaic technology to the railway trains: A case study for Istanbul airport-M1 light metro line', *17th IEEE International Conference on Smart Technologies, EUROCON 2017 – Conference Proceedings*, pp. 336–340. Available at: https://doi.org/10.1109/EUROCON.2017.8011131.

Kılkış, B. and Kılkış, Ş. (2017) 'New exergy metrics for energy, environment, and economy nexus and optimum design model for nearly-zero exergy airport (nZEXAP) systems', *Energy*, 140, pp. 1329–1349. Available at: https://doi.org/10.1016/J.ENERGY.2017.04.129.

Morganti, M. *et al.* (2019) 'Buildingmass and Energy Demand in Conventional Housing Typologies of the Mediterranean City', *Sustainability 2019, Vol. 11, Page 3540*, 11(13), p. 3540. Available at: https://doi.org/10.3390/SU11133540.

Rodríguez-Sanz, Á. *et al.* (2020) 'Airports and environmental sustainability: a comprehensive review', *Environmental Research Letters*, 15(10), p. 103007. Available at: https://doi.org/10.1088/1748-9326/ABB42A.

Sreenath, S., Sudhakar, K. and Yusop, A.F. (2020) 'Airport-based photovoltaic applications', *Progress in Photovoltaics: Research and Applications*, 28(8), pp. 833–853. Available at: https://doi.org/10.1002/PIP.3265.

Su, P.A. and Karney, B. (2014) 'Micro hydroelectric energy recovery in municipal water systems: A case study for Vancouver', *Urban Water Journal*, 12(8), pp. 678–690. Available at: https://doi.org/10.1080/1573062X.2014.923919.

Sukumaran, S. and Sudhakar, K. (2017) 'Fully solar powered airport: A case study of Cochin International airport', *Journal of Air Transport Management*, 62, pp. 176–188. Available at: https://doi.org/10.1016/J.JAIRTRAMAN.2017.04.004.

Zhou, Y. (2022) 'Low-carbon transition in smart city with sustainable airport energy ecosystems and hydrogen-based renewable-grid-storage-flexibility', *Energy Reviews*, 1(1), p. 100001. Available at: https://doi.org/10.1016/J.ENREV.2022.100001.

Zu, F. and Zhang, X. (2023) 'Cost-benefit analysis of micro WtE incinerator applications in an urban environment: A case study at the Hong Kong International Airport', *Waste Management*, 159, pp. 134–145. Available at: https://doi.org/10.1016/J.WASMAN.2023.01.032.

# Chapter 22
# Adaptation of Smart Energy Map to Transportation Domain: A Case Study of Small Airfield Buildings and Other Infrastructures

**Omar Alharasees, Utku Kale, Jozsef Rohacs, and Daniel Rohacs**

## 22.1 Introduction

The aviation industry's vital global role in transportation necessitates addressing its environmental impact, notably carbon emissions and energy consumption, for sustainability (Lai et al. 2022). This chapter analyzes sustainable energy for small airfield buildings, using the analytic hierarchy process to assess feasibility, environmental sustainability, and practical applications.

Waste-to-energy, as demonstrated at Stockholm Arlanda Airport, holds substantial promise in airfields (Kokkinos et al. 2023). It efficiently converts organic waste to heat or electricity through incineration or anaerobic digestion, providing waste management and energy generation solutions. Real-life cases, like the biogas plant at Stockholm Arlanda Airport, showcasing organic waste conversion to biogas for fuel, affirm the technology's practicality and environmental benefits ('Linde to build its first hydrogen fueling station in Sweden' 2015).

Biomass gasification is recognized for converting materials like agricultural residues and wood chips into synthesis gas (syngas) for power or biofuels (Rauch et al. 2014). Sweden's Gothenburg Biomass Gasification Project successfully produces clean syngas, reducing fossil fuel dependence (Thunman and Seemann 2019). Biofuel production from biomass offers promising sustainable energy for airfields

O. Alharasees (✉)
Department of Aeronautics, and Naval Architecture, Faculty of Transportation Engineering and Vehicle Engineering, Budapest University of Technology and Economics, Budapest, Hungary
e-mail: oalharasees@edu.bme.hu

U. Kale · J. Rohacs · D. Rohacs
Centre for Circular Economy Analysis and Knowledge at the Hungarian University of Agriculture and Life Sciences, Godollo, Hungary
e-mail: kale.utku@kjk.bme.hu; rohacs.jozsef@kjk.bme.hu; rohacs.daniel@kjk.bme.hu

 
T. H. Karakoc et al. (eds.), *The Future of Electric Aviation and Artificial Intelligence*, Sustainable Aviation, https://doi.org/10.1007/978-3-031-62094-2_22

and aerodromes (Baxter et al. 2020). Converting organic matter like crops or algae into oils for liquid or gaseous biofuels provides an eco-friendly alternative to fossil fuels. Seattle-Tacoma Airport's biofuel program, utilizing locally sourced waste cooking oil and forest residuals, showcases biomass biofuel's feasibility and potential for aviation adoption (Baxter et al. 2020). Besides waste-to-energy and biomass, solar power is a prominent player in aviation's shift to sustainability (Olabi and Abdelkareem 2022). Using photovoltaic panels, airports can generate clean, on-site electricity, decreasing grid reliance and carbon emissions. Cochin International Airport in India operates solely on solar power, exemplifying solar energy's vast potential in aviation (Sukumaran and Sudhakar 2017). Hydrogen technology provides promising sustainable energy for airfields and aerodromes. Fuel cells use hydrogen and oxygen to produce emissions-free electricity. Incheon Airport, South Korea, showcases this with hydrogen fuel cell buses, reducing carbon emissions and enhancing air quality ('Air Liquide completes hydrogen station for Incheon airport buses', 2021).

The aviation industry recognizes the imperative for sustainable energy in airfields and aerodromes. Diverse energy types such as waste-to-energy, biomass gasification, solar power, biomass-derived biofuels, and hydrogen technology offer enhanced efficiency, reduced emissions, and environmental gains. Using the analytic hierarchy process (AHP) methodology (Omar and Utku 2022), this research aims to comprehensively assess these energy types, supporting informed decisions and fostering sustainability in aviation. This study evaluates energy types for airfields and aerodromes using AHP and real-life cases, including waste-to-energy, biomass gasification, solar power, biomass-derived biofuels, and hydrogen technology. The results inform decisions and promote sustainable energy adoption, enhancing efficiency, cutting emissions, and advancing aviation sustainability.

## 22.2 Method

Sustainable energy is vital for airfield and aerodrome efficiency and environmental impact reduction. Using analytic hierarchy process (AHP) effective tool in sustainable aviation (Alharasees and Kale 2023), this study identifies promising energy types for micro-energetic systems in these environments, aiming for eco-friendly and effective operations. Initially, a tailored list of 12 sustainable energy types was compiled for airfield and aerodrome contexts, including

1. Waste-to-energy: Conversion of organic waste to heat or electricity. It was implemented at Stockholm Arlanda Airport for biogas fuel production.
2. Biomass gasification: Conversion of biomass to syngas for electricity or biofuel. Demonstrated by the successful Gothenburg Biomass Gasification Project.
3. Biofuel production from biomass: Extraction of oils from organic matter for sustainable aviation biofuel. Implemented at Seattle-Tacoma Airport using waste cooking oil and forest residuals.

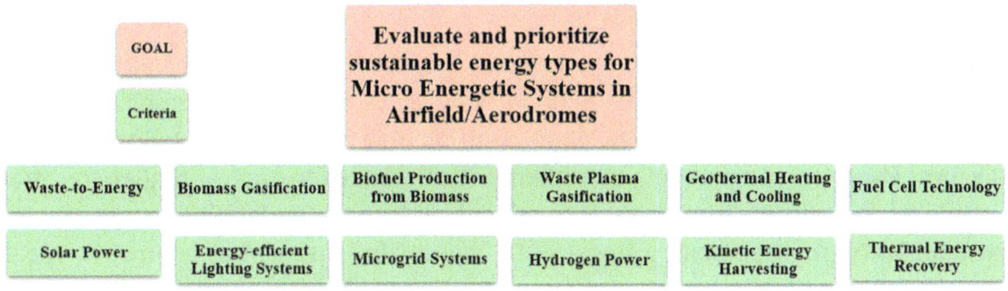


**Fig. 22.1** Airfield micro-energetic systems hierarchy

4. Waste plasma gasification: Conversion of waste to clean syngas using plasma arc (Sanjaya and Abbas 2023). Operated at Edmonton International Airport for municipal solid waste management.
5. Geothermal heating and cooling: Utilization of Earth's heat for airfield facility climate control. Utilized at Reykjavik Airport (Sætórsdóttir and Hall 2019), tapping into geothermal resources.
6. Fuel cell technology: Electricity generation using hydrogen and oxygen electrochemical reactions. It was introduced at Incheon Airport through hydrogen fuel cell buses.
7. Solar power: Electricity generation from sunlight through photovoltaic panels. Implemented at Cochin International Airport, meeting all energy demands.
8. Energy-efficient lighting systems: Adoption of LED technology for significant energy savings and improved lighting quality (Budd et al. 2015). Executed at London Heathrow Airport.
9. Microgrid systems: Integration of multiple energy sources and storage for self-sustaining power networks (Lidula and Rajapakse 2011). Implemented at Gibraltar International Airport for enhanced energy resilience.
10. Hydrogen power: Utilization of hydrogen fuel cells for clean electricity generation (Solvoll and Hanssen 2022). Piloted at Östersund Airport, showcasing hydrogen-powered snowplows and vehicles.
11. Kinetic energy harvesting: Conversion of mechanical motion to electrical energy (Correia and Ferreira 2021). Deployed at Heathrow Airport, harnessing passenger footsteps for low-energy devices.
12. Thermal energy recovery: Capture of waste heat for efficient energy utilization (Rasku and Kiviluoma 2018). Implemented at Helsinki Airport, utilizing data center waste heat for building heating.

The AHP assessment process involved assigning weights and ranking the energy types. The assessment used the "Saaty Scale" (Saaty 1977). Figure 22.1 illustrates the hierarchal model of assessment showing the investigated energy types. This study comprehensively assesses sustainable energy for airfields and aerodromes using AHP. It examines 12 energy types and real-life implementations, aiming to understand their benefits and implications, contributing to efficiency, emissions reduction, and aviation sustainability.

## 22.3 Results and Discussion

The AHP analysis revealed distinctions between expert groups, emphasizing the methodology's importance in evaluating sustainable energy for airfields and aerodromes. Pairwise comparisons identified critical characteristics, aiding energy type selection. Responses from engineering and architects, and building energy and sustainability management groups were aggregated using the geometric mean. Results reflected differing perspectives, offering insights for prioritizing sustainable energy solutions. Table 22.1 displays priority weights for energy types, as ranked by both expert groups, revealing the criteria's relative importance in the decision-making process.

The analysis shows solar power as the preferred energy type for both expert groups, with high weights and top rankings, emphasizing its recognition as a reliable and sustainable energy source for airfields and aerodromes. The consensus underscores the emphasis on renewable energy. Microgrid systems also received favorable rankings from both groups, indicating their perceived importance for energy resilience and source integration. The high rankings reflect the growing recognition of microgrids for enhancing energy reliability and efficiency in these environments. The variation reflects differing perspectives on technology suitability and benefits. These disparities indicate the necessity for further evaluation and consensus-building regarding energy types in these environments. Figure 22.2 visually depicts the comparison between group rankings, highlighting variations and underscoring the need for comprehensive assessment approaches like the analytic hierarchy process (AHP) to address divergent opinions.

The results of this analysis offer an in-depth vision for the selection and prioritization of sustainable energy solutions in airfield and aerodrome settings. They

**Table 22.1** AHP comparison weights and priority ranking for experts' groups

| | Engineering and architects' group | | Building energy and sustainability management group | |
|---|---|---|---|---|
| Criteria | Weights (%) | Rank | Weights (%) | Rank |
| Waste-to-energy | 2.25 | 11 | 6.30 | 6 |
| Biomass gasification | 4.30 | 7 | 2.30 | 12 |
| Biofuel production from biomass | 6.20 | 4 | 3.10 | 11 |
| Waste plasma gasification | 3.10 | 9 | 3.50 | 10 |
| Geothermal heating and cooling | 2.35 | 10 | 3.90 | 8 |
| Fuel cell technology | 5.90 | 5 | 11.50 | 4 |
| Solar power | 24.20 | 1 | 16.30 | 2 |
| Energy-efficient lighting systems | 3.95 | 8 | 9.00 | 5 |
| Microgrid systems | 17.44 | 3 | 14.90 | 3 |
| Hydrogen power | 23.7 | 2 | 19.8 | 1 |
| Kinetic energy harvesting | 4.71 | 6 | 5.60 | 7 |
| Thermal energy recovery | 1.90 | 12 | 3.80 | 9 |

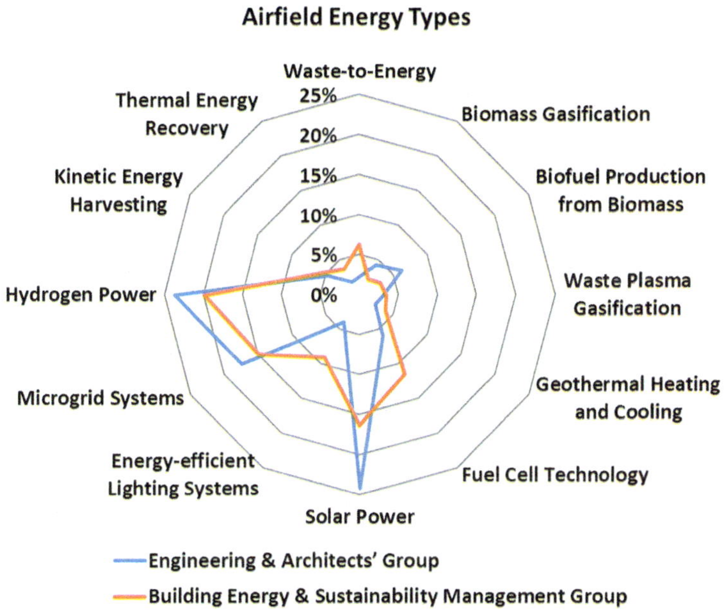


**Fig. 22.2** Airfield energy types: AHP criteria comparison

highlight the importance of considering multiple perspectives and stakeholder input to ensure a comprehensive evaluation of energy options and facilitate informed decision-making in pursuit of energy efficiency and environmental sustainability.

## 22.4 Conclusion

This chapter comprehensively assessed sustainable energy options suitable for small airfield buildings and other infrastructures. By employing the analytic hierarchy process (AHP) methodology, this study has shed light on the real-life implementation of various energy types in these settings.

Results favor solar power as the most preferred energy type, backed by high rankings and weights from expert groups. The growing recognition of solar power underscores its importance as a reliable and sustainable energy source. Microgrid systems also receive favorable rankings, highlighting their perceived significance for energy resilience and source integration. Prioritizing these energy types signifies the industry's dedication to efficiency and environmental impact reduction. However, divergent views persist regarding certain energy types, like fuel cell technology. This underscores the necessity for industry-wide evaluation and consensus-building to identify effective energy solutions for airfields and aerodromes. Comprehensive assessment methods like AHP offer a valuable framework for capturing and evaluating diverse opinions, aiding informed decisions and promoting sustainable energy adoption.

This chapter adds to the sustainable energy discourse in aviation by providing insights into suitable energy options for airfields and aerodromes. Prioritizing solar power and microgrid systems reflects industry commitment to renewable energy and resilience. The results inform the smart energy map, optimizing energy source combinations. Further research is necessary to address differing perspectives and ensure comprehensive evaluations. Embracing sustainable energy enhances efficiency, cuts emissions, and advances aviation's greener and sustainable future.

**Acknowledgments** This work was supported by Project No. 2022–2.1.1-NL-2022-00012, which has been implemented with the support provided by the Ministry of Culture and Innovation of Hungary from the National Research, Development and Innovation Fund, financed under the National Laboratories funding scheme.

## References

'Air Liquide completes hydrogen station for Incheon airport buses' (2021) *Fuel Cells Bulletin*, 2021(9), pp. 10–11. Available at: https://doi.org/10.1016/S1464-2859(21)00510-1.

Alharasees, O. and Kale, U. (2023) 'Applying AHP for supplier selection in aviation: a multi-criteria decision-making approach', *International Journal of Sustainable Aviation*, 9(4), pp. 293–313. Available at: https://doi.org/10.1504/IJSA.2023.134344.

Baxter, G., Srisaeng, P. and Wild, G. (2020) 'The Use of Aviation Biofuels as an Airport Environmental Sustainability Measure: The Case of Oslo Gardermoen Airport', *MAD – Magazine of Aviation Development*, 8(1), pp. 6–17. Available at: https://doi.org/10.14311/MAD.2020.01.01.

Budd, T., Budd, L. and Ison, S. (2015) 'Environmentally sustainable practices at UK airports', *Proceedings of the Institution of Civil Engineers – Transport*, 168(2), pp. 116–123. Available at: https://doi.org/10.1680/TRAN.13.00076.

Correia, D. and Ferreira, A. (2021) 'Energy Harvesting on Airport Pavements: State-of-the-Art', *Sustainability 2021, Vol. 13, Page 5893*, 13(11), p. 5893. Available at: https://doi.org/10.3390/SU13115893.

Kokkinos, Nikolaos C *et al.* (2023) 'Waste-to-Energy: Applications and Perspectives on Sustainable Aviation Fuel Production', pp. 265–286. Available at: https://doi.org/10.1007/978-981-99-1392-3_10.

Lai, Y.Y. *et al.* (2022) 'Analysing the opportunities and challenges for mitigating the climate impact of aviation: A narrative review', *Renewable and Sustainable Energy Reviews*, 156, p. 111972. Available at: https://doi.org/10.1016/J.RSER.2021.111972.

Lidula, N.W.A. and Rajapakse, A.D. (2011) 'Microgrids research: A review of experimental microgrids and test systems', *Renewable and Sustainable Energy Reviews*, 15(1), pp. 186–202. Available at: https://doi.org/10.1016/J.RSER.2010.09.041.

'Linde to build its first hydrogen fueling station in Sweden' (2015) *Fuel Cells Bulletin*, 2015(2), pp. 7–8. Available at: https://doi.org/10.1016/S1464-2859(15)30019-5.

Olabi, A.G. and Abdelkareem, M.A. (2022) 'Renewable energy and climate change', *Renewable and Sustainable Energy Reviews*, 158, p. 112111. Available at: https://doi.org/10.1016/J.RSER.2022.112111.

Omar, A. and Utku, K. (2022) 'Air Transport Projects Quality Assessments by Analytical Hierarchy Process (AHP)', *Repüléstudományi Közlemények*, 34(2), pp. 73–82. Available at: https://doi.org/10.32560/RK.2022.2.6.

Rasku, T. and Kiviluoma, J. (2018) 'A Comparison of Widespread Flexible Residential Electric Heating and Energy Efficiency in a Future Nordic Power System', *Energies 2019, Vol. 12, Page 5*, 12(1), p. 5. Available at: https://doi.org/10.3390/EN12010005.

Rauch, R., Hrbek, J. and Hofbauer, H. (2014) 'Biomass gasification for synthesis gas production and applications of the syngas', *Wiley Interdisciplinary Reviews: Energy and Environment*, 3(4), pp. 343–362. Available at: https://doi.org/10.1002/WENE.97.

Saaty, T.L. (1977) 'A scaling method for priorities in hierarchical structures', *Journal of Mathematical Psychology*, 15(3). Available at: https://doi.org/10.1016/0022-2496(77)90033-5.

Sætórsdóttir, A.D. and Hall, C.M. (2019) 'Contested Development Paths and Rural communities: Sustainable Energy or Sustainable Tourism in Iceland?', *Sustainability 2019, Vol. 11, Page 3642*, 11(13), p. 3642. Available at: https://doi.org/10.3390/SU11133642.

Sanjaya, E. and Abbas, A. (2023) 'Plasma gasification as an alternative energy-from-waste (EFW) technology for the circular economy: An environmental review', *Resources, Conservation and Recycling*, 189, p. 106730. Available at: https://doi.org/10.1016/J.RESCONREC.2022.106730.

Solvoll, G. and Hanssen, T.-E.S. (2022) 'Public Service Obligation as a tool for implementing flight routes operated by electric aircrafts', *86* [Preprint]. Available at: https://nordopen.nord.no/nord-xmlui/handle/11250/3013229 (Accessed: 12 June 2023).

Sukumaran, S. and Sudhakar, K. (2017) 'Fully solar powered airport: A case study of Cochin International airport', *Journal of Air Transport Management*, 62, pp. 176–188. Available at: https://doi.org/10.1016/J.JAIRTRAMAN.2017.04.004.

Thunman, H. and Seemann, M. (2019) 'The GoBiGas plant', *Substitute Natural Gas from Waste: Technical Assessment and Industrial Applications of Biochemical and Thermochemical Processes*, pp. 455–474. Available at: https://doi.org/10.1016/B978-0-12-815554-7.00017-9.

# Chapter 23
# Modeling the Noise Characteristics of a Regional Turboprop Hybrid Electric Aircraft

**Oleksandr Zaporozhets, Vitalii Makarenko, Vadim Tokarev, Kateryna Kazhan, and Kateryna Synylo**

## Nomenclature

| | |
|---|---|
| AN | Aviation noise |
| EPNL | Effective perceived noise level |
| HEA | Hybrid electric aircraft |
| LAmax | Maximum A-weighted sound pressure level |
| PP | Power plant |
| SEL | Sound exposure level |

## 23.1 Introduction

In this chapter, we evaluate the possibility of reducing the noise of a regional hybrid-electric aircraft (HEA) with a parallel electric drive, in which the gas turbine and electric engines are connected to the propeller through a common gearbox (Shmyrov et al. 2020). As an example, calculations were made for the modification of the An-26 regional aircraft model. The efficiency of HEA operation was evaluated using criteria including fuel consumption and local noise characteristics. Calculations of the parameters of power plants (PPs) and the HEA trajectory were performed. We

O. Zaporozhets
Łukasiewicz Research Network – Institute of Aviation, Warsaw, Poland
e-mail: oleksandr.zaporozhets@ilot.lukasiewicz.gov.pl

V. Makarenko (✉) · V. Tokarev · K. Kazhan · K. Synylo
National Aviation University, Kyiv, Ukraine
e-mail: vitmakarenko@nau.edu.ua; tokarev@nau.edu.ua; kazhank@nau.edu.ua; synylo@nau.edu.ua



T. H. Karakoc et al. (eds.), *The Future of Electric Aviation and Artificial Intelligence*, Sustainable Aviation, https://doi.org/10.1007/978-3-031-62094-2_23

demonstrate the obtained acoustic characteristics of the HEA for a typical flight profile.

## 23.2 Modeling

The flight path is the basis on which the HEA noise calculation is performed. The parameters of the HEA trajectory were determined using differential equations:

$$\begin{cases} \frac{dv}{dt} = \frac{1}{m}(T\cos\varepsilon - D - W\sin\theta), \\ \frac{d\theta}{dt} = \frac{1}{mv}(T\sin\varepsilon + L - W\cos\theta), \\ \frac{dh}{dt} = v\sin\theta, \\ \frac{dx}{dt} = v\cos\theta, \\ \frac{dm}{dt} = -G, \end{cases}$$

where $v$ is the flight speed, $t$ is the flight time, $m$ is the mass of the HEA at time $t$, $T$ is the total thrust of PP (for the HEA model, the TV3-117VMA-SBM1 turboshaft engine is used, but with reduced power and mass by 90%), $\varepsilon = \alpha + \beta$, $\alpha$ is the angle of attack, $\beta$ is the angle of installation of the engines in relation to the wing chord, $D$ is the resistance of the HEA, $W = mg$ is the weight of the HEA, $\theta$ is the trajectory angle, $L$ is the lifting force, $h$ is the flight height, $x$ is the length of the projection of the flight profile on the horizontal plane from the beginning of the flight segment, and $G$ is the total fuel consumption at time $t$.

The flight profile was calculated for an airport located at sea level with standard atmosphere parameters. The solution of the system of equations was found by the Runge–Kutta numerical method (MATLAB function “ode45”), where the values of the variables $v$, $\theta$, $h$, $m$ from the end of the previous flight segment were used as initial values. The initial value of the variable $x$ was always assumed to be zero. On the first segment of the flight, the initial values of the variables $x$, $v$, $\theta$, $h$ are assumed to be zero, except for the mass $m$, which is assumed to be equal to the maximum takeoff mass of the HEA. The flight profile of HEA must comply with the flight operations manual of An-26, except for the rates of climb and descent. In order to meet the limitations specified in the flight operation manual, optimization of pilot inputs using genetic algorithm was done. Based on the simulation results, a comparison of the profiles for the An-26 aircraft using the conventional PP (AI-24 + RU-19-300 with a four-bladed propeller) and a hybrid PP (turboshaft engine connected to a six-bladed propeller through a gearbox together with 900 kW electric motor) was performed (Fig. 23.1). The electric motor is used starting from the beginning of takeoff roll at 100% power for the first 1.5 min, then at 72% for the next 1.5 min, and

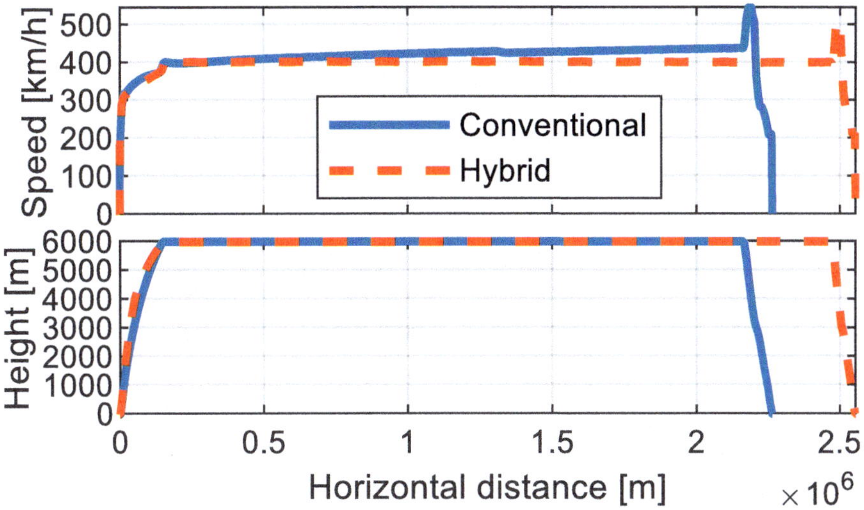


**Fig. 23.1** Comparison of flight profiles of an aircraft with a conventional and hybrid PP

after that, there was a linear reduction of power from 72% to zero for 9 min. It is planned to use storage batteries with a specific capacity of 435 Wh/kg with a total weight of 600 kg. The total weight of the hybrid PP is 566 kg more compared to the conventional PP, so the fuel weight is reduced by this value in the calculations.

## 23.3 Results

Simulation shows that the HEA with the same payload (2.7 tons) flew 290.9 km further compared to the An-26 with a conventional PP. However, this result may be due to the fact that the hybrid PP included a six-bladed propeller with a higher efficiency compared to the four-bladed propeller used in the conventional PP. Figure 23.2 shows that the hybrid aircraft consumes less fuel per unit of flight time. The results of modeling the levels of aviation noise (AN) according to the sound exposure level (SEL), maximum A-weighted sound pressure level (LAmax), and effective perceived noise level (EPNL) criteria at the certification points are presented in Table 23.1. From the calculation of results given in Table 23.1, it follows that a small reduction in noise is observed only at control points 2 and 3, in which the takeoff phase has the predominant contribution. Contours of equal noise were constructed and a comparison of the shape of the contours was carried out according to the LAmax, SEL, and EPNL criteria for the landing and takeoff cycle of the calculated type of aircraft (An-26). The areas of equal noise contours are given in Table 23.2.

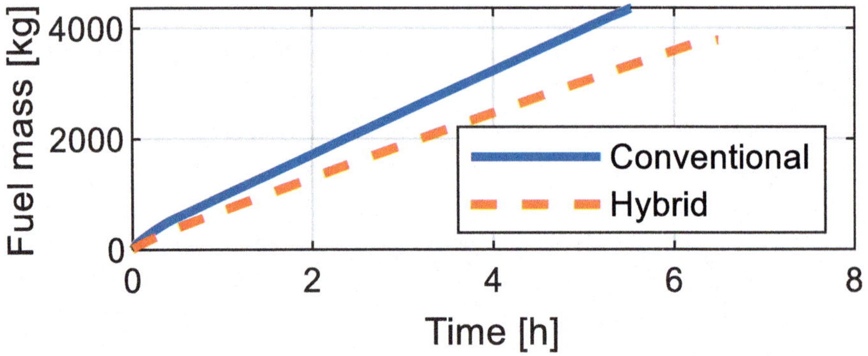


**Fig. 23.2** Fuel consumption during the flight

**Table 23.1** AN sound levels according to the SEL, LAmax, and EPNL criteria at certification points based on simulation results for an aircraft with conventional and hybrid PP and comparison with the results of certification noise measurements for An-26 aircraft

| Certification point number | | 1 | 2 | 3 |
|---|---|---|---|---|
| Coordinates | X, м | 2000 | 1754 | 6500 |
| | Y, м | 0 | 450 | 0 |
| SEL | Conventional | 95.6 | 86.5 | 85.4 |
| | Hybrid | 95.6 | 86.8 | 84.9 |
| LAmax | Conventional | 89.6 | 76.6 | 76.1 |
| | Hybrid | 89.6 | 76.6 | 75.1 |
| EPNL | Conventional | 103.4 | 92.5 | 91.4 |
| | Hybrid | 103.4 | 92.3 | 90.7 |
| EPNL measured at certification of An-26 | | **100.8** | **92.6** | **91.1** |
| Difference | Conventional | 2.6 | −0.1 | 0.3 |
| | Hybrid | 2.6 | −0.3 | −0.4 |

**Table 23.2** Comparison of equal noise contour areas for conventional An-26 and HEA

| | EPNL | | LAmax | | SEL | |
|---|---|---|---|---|---|---|
| | Conven. | HEA | Conven. | HEA | Conven. | HEA |
| dB/EPNLdB | $km^2$ | $km^2$ | $km^2$ | $km^2$ | $km^2$ | $km^2$ |
| 65 | – | – | 29.207 | 28.264 | – | – |
| 70 | 153.907 | 145.889 | 15.152 | 14.309 | 116.705 | 115.836 |
| 75 | 82.703 | 80.666 | 7.529 | 7.302 | 48.856 | 50.579 |
| 80 | 40.864 | 40.153 | 3.655 | 3.685 | 18.234 | 18.87 |
| 85 | 19.281 | 18.973 | 1.717 | 1.781 | 7.555 | 7.671 |
| 90 | 9.426 | 9.168 | 0.816 | 0.829 | 2.978 | 3.156 |
| 95 | 4.125 | 4.214 | 0.367 | 0.362 | 1.171 | 1.213 |
| 100 | 1.756 | 1.791 | 0.158 | 0.153 | 0.448 | 0.454 |
| 105 | 0.698 | 0.697 | 0.075 | 0.076 | 0.166 | 0.171 |

## 23.4 Conclusion

Thus, for the given initial conditions according to the methods of International Civil Aviation Organization and European Civil Aviation Conference, the assessment of the levels of AN in the certification points was performed and contours of equal noise were constructed according to the most important criteria for single noise events: LAmax, SEL, EPNL.

A comparison of the simulation results according to the EPNL criterion with the certification data at the control points showed that the deviations do not exceed 3 EPNLdB. This confirms the further possibility of using the developed AN model to study the impact of different degrees of HEA hybridization. As well as to use it in the study of the impact of technological solutions on the general noise levels in the vicinity of civil aviation airports for various scenarios of operation of the HEA fleet, as well as when solving optimization problems.

## Reference

Shmyrov V, Loginov V, Fil S, Khaustov A, Bondarchuk O, Kalashnikov A, Khmelnitskiy G (2020) The modernization concept of aircraft An-26 and An-140 based on the use of a hybrid power system. Eastern-European Journal of Enterprise Technologies. 5(1):6-17. https://doi.org/10.15587/1729-4061.2020.212150

# Chapter 24
# Concepts of Commercial Aircraft with Hybrid Propulsion

**Paweł Abratowski**

## Nomenclature

| | |
|---|---|
| BLI | Boundary layer ingestion |
| FEM | Finite element method |
| HEP | Hybrid electric propulsion |
| IMOTHEP | Investigation and Maturation of Technologies for Hybrid Electric Propulsion |
| SMR | Small/medium range |

## 24.1 Introduction

The main goal of the Investigation and Maturation of Technologies for Hybrid Electric Propulsion (IMOTHEP) project is to assess the potential offered by hybrid electric propulsion (HEP). The four aircraft concepts focused on regional and short-to-medium-range aircraft missions are presented in the chapter. At the end of the chapter, the wing structure strength calculation samples are also presented for the aircraft regional configuration (Fig. 24.1).

P. Abratowski (✉)
Łukasiewicz Research Network, Institute of Aviation, Warsaw, Poland
e-mail: pawel.abratowski@ilot.lukasiewicz.gov.pl



T. H. Karakoc et al. (eds.), *The Future of Electric Aviation and Artificial Intelligence*, Sustainable Aviation, https://doi.org/10.1007/978-3-031-62094-2_24

| | Conservative | Radical |
|---|---|---|
| Regional |  Assisted turboshaft |  Turboelectric with DEP and wing-tip propeller |
| SMR |  Turboelectric with DEP |  Turboelectric BWB with DEP and BLI |

**Fig. 24.1** Four configurations of hybrid aircraft. (Defoort 2020)

**Fig. 24.2** SMR conservative configuration. (Defoort 2020)

## 24.2 The "SMR Conservative" Concept

The "SMR conservative" (short-medium-range conservative) configuration is based on Airbus A320. The visualization of the aircraft is shown in Fig. 24.2. The propulsion consists of distributed 24 electric motors driving fans. Two turbine engines drive two generators. The fans are built in a way that allows for the use of the boundary layer ingestion effect (BLI).

## 24.3 The "SMR Radical Concept"

The aircraft is a new design in tailless configuration. The fuselage works partly as a wing in this configuration. The propulsion consists of two gas turbines and eight electric motors driving fans. All are placed at the back part of a fuselage. Additionally, as in "SMR conservative," BLI effect is used. The aircraft has two gas turbines driving two generators. The visualization of the aircraft is shown in Fig. 24.3.

## 24.4 The "Regional Conservative" Concept

The "regional conservative" (regional range conservative configuration) is based on ATR-42-600. The propulsion is integrated parallel hybrid power plant system. A fan is driven directly by a gas turbine as well as an electric motor. The gas turbine drives also an electric generator. Integrated parallel hybrid system is located in nacelle. The aircraft propulsion scheme is shown in Fig. 24.4.

## 24.5 The "Regional Radical" Concept

The "regional radical" (regional range radical configuration) is based on ATR-42-600. The propulsion has eight electric motors driving fans and powered by batteries. A turbine engine is placed in one nacelle (right side) and drives a generator. The second nacelle (left side) consists of battery pack for mass balance with the gas turbine (Atanasov 2020). The aircraft propulsion scheme is shown in Fig. 24.5.

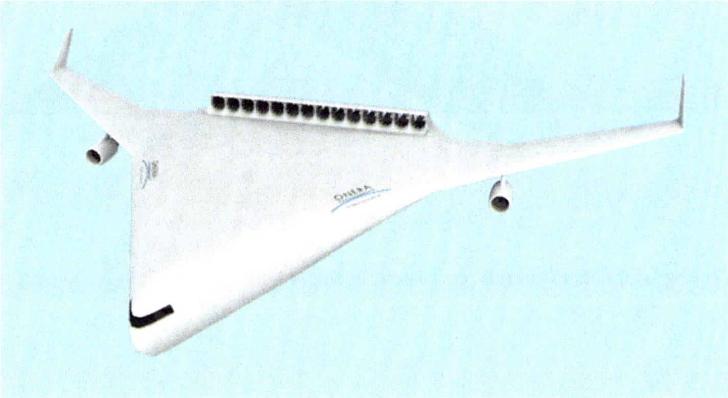

**Fig. 24.3** SMR radical aircraft configuration. (Defoort 2020)

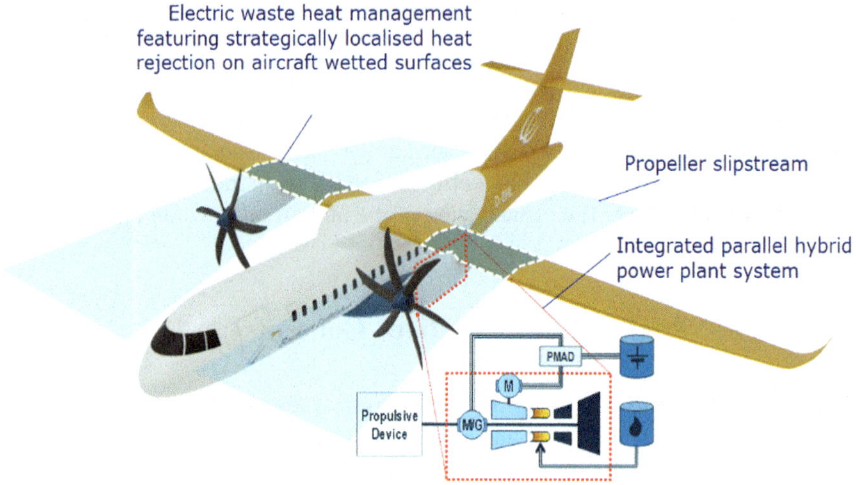


**Fig. 24.4** Regional conservative aircraft propulsion scheme. (Defoort 2020)

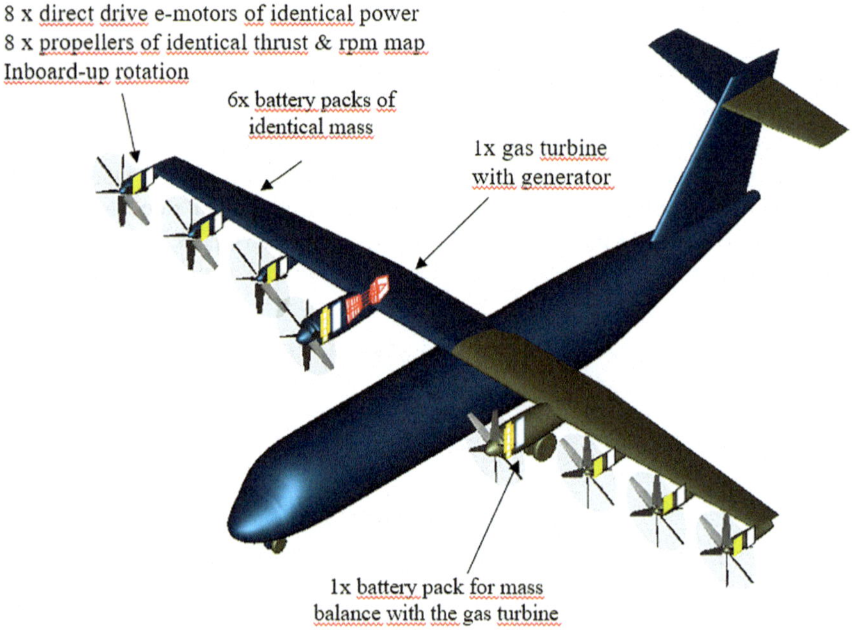


**Fig. 24.5** Regional radical aircraft propulsion scheme. (Author)

## 24.6 The Sample of the “Regional Radical” Wing Strength Calculations

The simplified finite element method (FEM) model of the wing has been created. Many simplifications have been applied—the electric motors, generators, batteries, etc., have been modeled as point masses. As a critical case, the “landing with braking” (CS25 regulations) has been applied. This is a suitable load case for a long wing loaded with heavy distributed masses. The main loads are created by inertia forces. The wing structure is made from CFRP composite. The scheme of the FEM model is shown in Fig. 24.6. The sample of FEM calculations result is shown in Fig. 24.7.

## 24.7 Concluding Remarks

Aircraft configurations will be selected based on their potential to reduce fuel consumption and their representativeness of a variety of credible concepts, with a focus on regional and short-to-medium-range missions.

The strength calculations have been performed for regional-radical configuration. This version is foreseen for further development most probably. The mass of the wing structure at this step of analysis is within the assumed range (at this stage). Other load cases are also assumed.

The results should be taken as the direction of the wing structure for further development. The FEM model also allows for modal analysis.

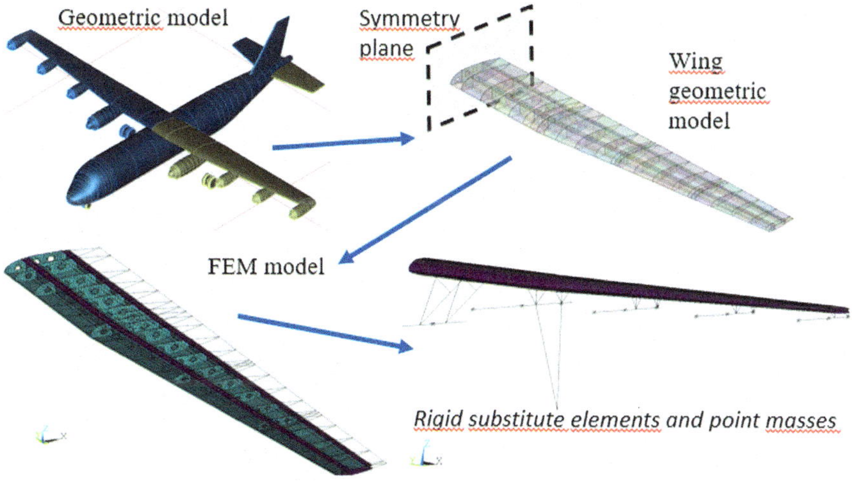


**Fig. 24.6** Regional radical aircraft propulsion scheme. (Author)

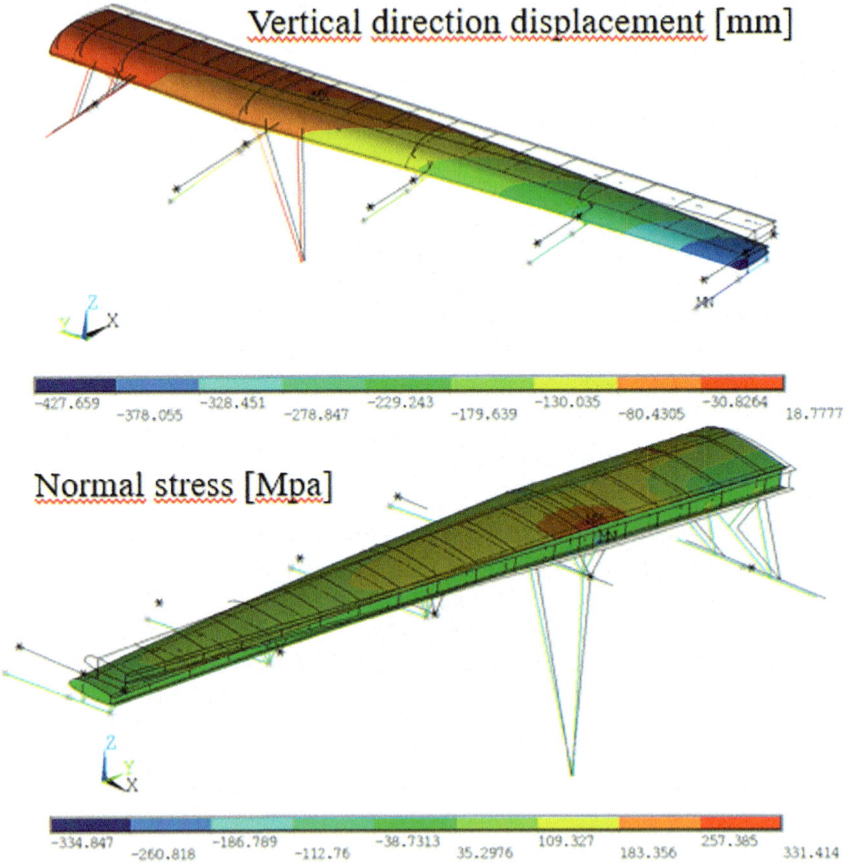


**Fig. 24.7** Regional radical aircraft propulsion scheme. (Author)

## References

G. Atanasov and DLR, Presentation: Preliminary Results on REG-RAD, 2020
S. Defoort (2020), Integrated Vehicle Design, Slide 9., Presentation IMOTHEP IMOTHEP https://www.imothep-project.eu/

# Chapter 25
# Highly Accurate and Confident Basic Aircraft Noise Assessment Scenario

**Andrzei Chyla, Michael Bukala, Kateryna Kazhan, Vitalii Makarenko, Vadim Tokarev, and Olexandr Zaporozhets**

## 25.1 Introduction

Noise is and will be one of the noteworthy phenomena accompanying air transport. Taking into account the constant increase in air traffic, a number of attempts have been made to reduce the aviation noise exposure. The traditional approach to this problem is to identify and classify the main sources of aircraft noise and define their reduction measures (Zaporozhets et al. 2011). For this purpose, more stringent restrictions are introduced successively in a stepwise manner, based on the aircraft noise certification process. Noise performance assessment at the aircraft level will be done for the classes of aircraft by revisiting aircraft configurations and taking into account new onboard energy storage and management systems and efficient green operations. The flight management system will integrate new functions for trajectory optimization, adapted climb–cruise–approach profiles driven by airline constraints, weather effects, traffic en route, formation flight, or ground operation management to include zero-$CO_2$ pushback and taxiing phases, etc. The efficiency of new designs and flight management systems will be shown by comparison with existing airplane types (basic airplane flight noise scenario) in operation—for single-flight noise event first of all (Kazhan et al. 2023a). For higher confidence and accuracy of the assessment, the basic airplane flight noise scenario is assessed by measurements

A. Chyla (✉) · K. Kazhan · V. Makarenko · V. Tokarev
National Aviation University, Kyiv, Ukraine

M. Bukala
National Aviation University, Kyiv, Ukraine

Noise LLC, Warsaw, Poland

O. Zaporozhets
Łukasiewicz Research Network Institute of Aviation, Warsaw, Poland

 
T. H. Karakoc et al. (eds.), *The Future of Electric Aviation and Artificial Intelligence*, Sustainable Aviation, https://doi.org/10.1007/978-3-031-62094-2_25

(Zaporozhets et al. 2023) and calculation in similar conditions equal to the current regional airport (Kazhan et al. 2023b).

## 25.2 Basic Scenario

The basic scenario is defined for comparison of the new technologies' benefits with existing analog, which is the perspective for further development. The class of regional airplanes like ATR-72-500 and Dash-8-400 is considered as such a group for short-range connections by changing their power plant on electric energy consumption. The thrust of the electrified airplanes must be produced by propellers by its basic design, meaning the dominant acoustic source will also be the same. The efficiency of a new power plant should be analyzed from the point of view of concentrated or distributed power being used for thrust providing, but in any case the basic design is defined by the ATR-72 class. To date, the ANP database considered the turboprop Hawker Siddeley HS-748 with engines DART MK532-2 (HS748A) as an airplane for noise assessment of any other type from this group. The calculated noise contours for ATR-72-500 and Dash-8-400 are shown in Fig. 25.1.

All the types in consideration were certified for noise, and their levels at certification points 2 (takeoff) and 3 (approach) are shown in Table 25.1. The calculated LAmax for the same conditions is different at 1–2 dBA for HS748A and DH8300,

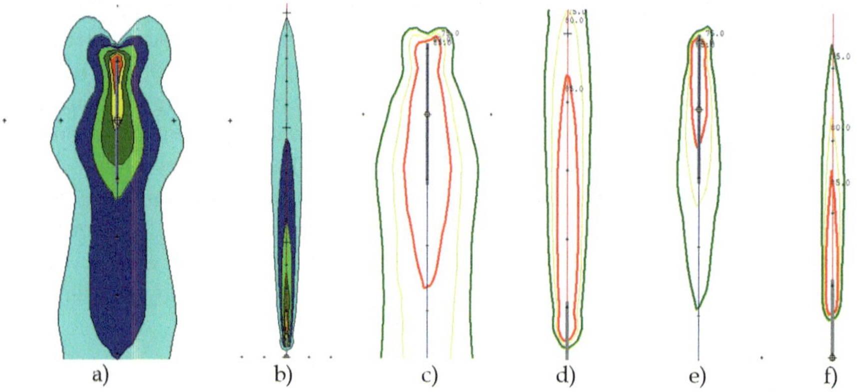


**Fig. 25.1** *SEL* contours for Dash-8-400 (**a**, **b**) and ATR-72-500 (**c**, **d**) at takeoff and approach, and the $L_{\mathrm{Amax}}$ contours for ATR-72-500 (**e**, **f**) for comparison with *SEL*

**Table 25.1** Certified $L_{\mathrm{Amax}}$ dBA/calculated by INM 6.1

| ANP identificator | Airplane type description | Takeoff | Approach |
|---|---|---|---|
| HS-748A | BAE-748 Series 2A RR DART MK532-2L | 78.3/76.2 | 88.8/87.1 |
| DH8300 | DHC-8 314 PW123 | 67.1/67.4 | 81.2/78.4 |
| ATR72 | ATR72-210 PW127 | 71.8/76.2 | 82.2/87.1 |

but the difference is quite huge for the ATR72. The real ATR-72 was included in 2023 in the ANP database (https://www.easa.europa.eu/.../aircraft-noise-and-performance-anp-data) for aircraft noise calculations. The exposure levels for this class of airplanes are higher at 9.5 dBA for takeoff and 7.5 dBA for approach noise.

HS748A is quite old and its noise performances were defined by technology that is 20–30 years old, some even older. Its certified noise levels are 7–10 dBA higher than ATR72 and DH8300 (Table 25.1). Also, flight performances were changed considerably to be more energy-efficient and have a lesser impact on the flight environment. Dash-8-400 should be calculated by substitution of Dash-8-300, which is very similar in flight and noise performances.

## 25.3 Noise Measuring and Calculations of Turboprop Aircraft

Aircraft noise assessment scenario should be established for referencing results to the newly designed electric aircraft according to the needs of the EU EFACA project (https://efaca.eu/). It must be accurate for a single flight for the selected turboprop plane. For regional turboprop airplane, the prototype is combined by the assessment of ATR-72 and Dash-8 in operation at a few regional airports in Poland (Fig. 25.2). The average takeoff and landing profiles were calculated from the recorded ADS-B takeoff and landing profiles (Fig. 25.3). The two actual routes/profiles were selected that were closest to the average of the aircraft under consideration. Noise measurements were done by continuous noise monitoring systems along with ADS-B data processing and analysis for more adequate flight performance assessment during the departure and arrival flight stages (Fig. 25.4). Evidently, the approaches were found to be quite similar to those defined by ANP data. For departures, the differences were defined, particularly for altitudes along the distance from the runway start. The power settings were used similar to those defined in the ANP database. The difference should be valid for monitoring results also.

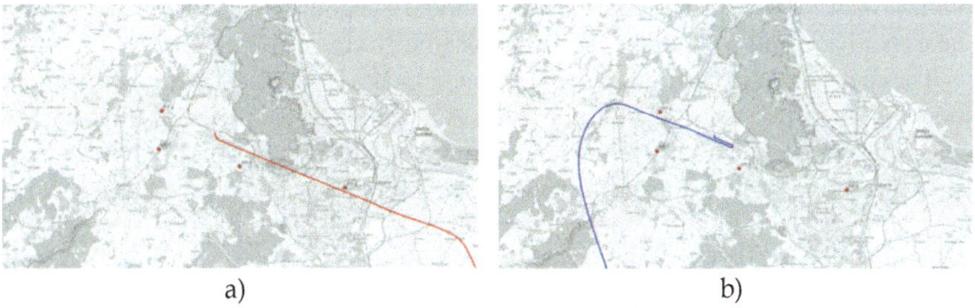


**Fig. 25.2** Flight routes for approach (**a**) and departure (**b**) at Gdansk airport: red points show the locations for noise monitoring

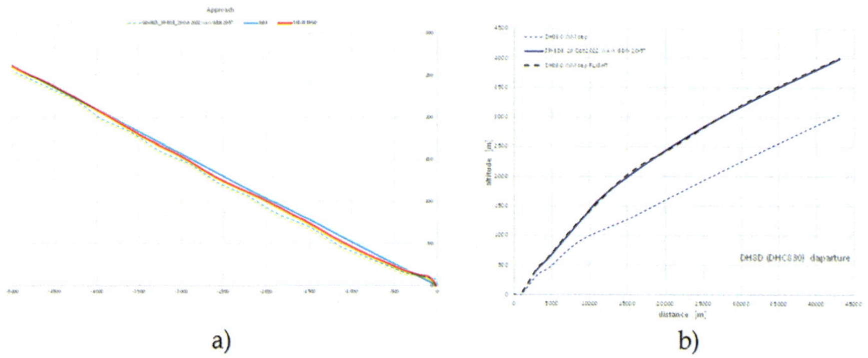

**Fig. 25.3** Flight profiles for approach (**a**) and departure (**b**) of Dash-8 at Gdansk airport: comparisons were done using the ANP database

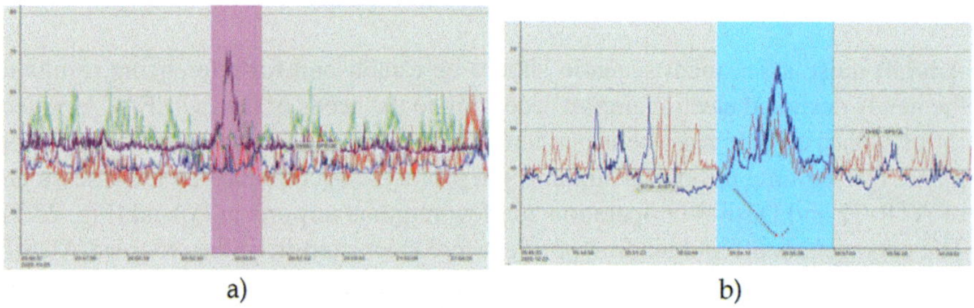

**Fig. 25.4** Flight noise *SEL* for approach (**a**) and departure (**b**) of Dash-8 at Gdansk airport: comparisons were done for different monitors

Aircraft noise exposure is calculated using the tools verified by International Civil Aviation Organization / Committee on Aviation Environmental Protection (ICAO/ CAEP) in accordance with the ICAO Doc 9911: IsoBella (aircraft noise exposure calculations, designed by the NAU, verified during the CAEP/13 working cycle) and AEDT 3.0 (a software system from the US FAA that models aircraft performance in space and time to estimate fuel consumption, emissions, noise, and air quality consequences at the regional, national, and global levels) with BADA (base of aircraft data, which provides theoretical model specifications and related specific datasets to accurately simulate the behavior of any aircraft).

The measured actual flight parameters (*x*, *y*, *z*, *V*, *T*) were used in computational models for aircraft noise level assessment. A graphical representation (Fig. 25.5) of *SEL* contours for the conventional configurations, in terms of noise exposure levels (dBA) during a standard takeoff/approach, highlights the potential concerns during the transition toward pollution-free aviation. Also, calculations made it possible to determine the *SEL* at the locations of the measurement points at the model airport.

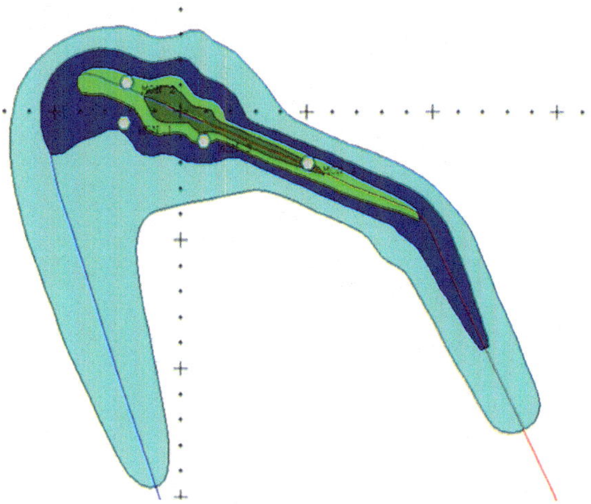

**Fig. 25.5** Flight noise *SEL* contours calculated for approach (**a**) and departure (**b**) of Dash-8 at Gdansk airport: noise monitors are shown by gray points

## 25.4 Conclusion

The differences between measured and calculated *SEL*s are quite huge for further consideration of Dash-8 as a basic airplane for comparison with new airplane with new electric power plants. Further investigation using a similar approach is proposed for ATR72 airplane, which is mostly appropriate for further development and implementation at regional air routes.

## References

Kazhan K., Makarenko V., Tokarev V., Synylo K., Zaporozhets O., Chyla A., Bukala M. (2023a) Noise assessment scenarios for new airplane climate efficient designs. This book. https://doi.org/10.1007/978-3-031-32639-4_11

Kazhan K., Zaporozhets O., Karpenko S. (2023b) Peculiarities of Pre-processing of ADS-B Data for Aircraft Noise Modeling and Measurement During Specific Stages of LTO Cycle In book: Advances in Electric Aviation (pp.83–89) https://doi.org/10.1007/978-3-031-32639-4_11

Zaporozhets O., V. Tokarev, K. Attenborough (2011) AIRCRAFT NOISE: assessment, prediction and control. – CRC Press Reference, 2011. – 420 p. (ISBN 9781138073029) https://doi.org/10.1201/b12545

Zaporozhets O., Gulevets V., Karpenko S., Kazhan K., Konovalova O., Paraschanov V. (2023) Aircraft Noise Measurements in Ukrainian Airports In book: Advances in Electric Aviation (pp.157–168) https://doi.org/10.1007/978-3-031-32639-4_21

# Index





T. H. Karakoc et al. (eds.), *The Future of Electric Aviation and Artificial Intelligence*, Sustainable Aviation, https://doi.org/10.1007/978-3-031-62094-2

SPRINGER NATURE

# GPSR Compliance

*The European Union's (EU) General Product Safety Regulation (GPSR) is a set of rules that requires consumer products to be safe and our obligations to ensure this.*

*If you have any concerns about our products, you can contact us on ProductSafety@springernature.com*

In case Publisher is established outside the EU, the EU authorized representative is:

Springer Nature Customer Service Center GmbH
Europaplatz 3
69115 Heidelberg, Germany

The manufacturer's authorised representative in the EU is Springer Nature Customer Service Centre GmbH, Europaplatz 3, 69115 Heidelberg, Germany. If you have any concerns regarding our products, please contact ProductSafety@springernature.com

Printed and bound by CPI Group (UK) Ltd, Croydon, CR0 4YY
30/09/2026
02225604-0011